DSP Processor Fundamentals

DSP Processor Fundamentals

Architectures and Features

Phil Lapsley
Jeff Bier
Amit Shoham
Berkeley Design Technology, Inc.

Edward A. Lee
University of California at Berkeley

The Institute of Electrical and Electronics Engineers, Inc., New York

A JOHN WILEY & SONS, INC., PUBLICATION
New York • Chichester • Weinheim • Brisbane • Singapore • Toronto

For ordering and customer service, call 1-800-CALL-WILEY.
Wiley-IEEE Press **ISBN 0-7803-3405-1**

This is the IEEE reprinting of a book previously published by Berkeley Design Technology, Inc., 39355 California Street, Suite 206, Fremont, CA 94538.
(510) 791-9100

ISBN 0-7803-3405-1

Library of Congress Cataloging-in-Publication Data

DSP processor fundamentals : architectures and features / Phil Lapsley
... [et al.].
Includes index.
ISBN 0-7803-3405-1 (pbk.)
1. Signal processing–Digital techniques–Equipment and supplies.
2. Microprocessors. I. Lapsley, Phil (date)
TK5102.9.D754 1997
621.382'2'0285416–dc20 96-41738
CIP

**To Svend Pedersen, Ira J. Meyer, Yair Shoham,
and Dave Messerschmitt**

Contents

Preface

This book is an introduction to the technical fundamentals of the architectures and features of programmable digital signal processors. Programmable digital signal processors (often called DSPs, PDSPs, or DSP processors) are microprocessors that are specialized to perform well in digital signal processing-intensive applications. Since the introduction of the first commercially successful DSP processors in the early 1980s, dozens of new processors have been developed, offering system designers a vast array of choices. According to the market research firm Forward Concepts, sales of user-programmable DSPs will total roughly US $1.8 billion in 1996, with a projected annual growth rate of 35 percent [Str95]. With semiconductor manufacturers vying for bigger shares of this booming market, designers' choices will broaden even further in the next few years.

Scope and Purpose

This book is intended for anyone who is evaluating or comparing DSP processors, designing DSP processors or systems employing DSP processors, or who wants an independent, comprehensive introduction to the technology. We present each of the key elements of DSP processor technology and examine current product offerings with a critical eye. We expect that this book will be especially useful for electronic systems designers, processor architects, engineering managers, and product planners. It will aid in choosing the DSP processor or processors that are best suited to a given application and in developing an understanding of how the capabilities of DSP processors can be used to meet the needs of the application. It is also intended to serve as an extensive tutorial and overview for those who wish to quickly master the essential aspects of DSPs.

Organization

This book is organized as follows:

- **Digital Signal Processing and DSP Systems**
  Chapter 1 provides a high-level overview of digital signal processing, including DSP system features and applications.
- **DSP Processor Embodiments and Alternatives**
  Chapter 2 provides a brief introduction to DSP processors and then discusses the different forms that DSP processors take, including chips, multichip modules, and cores. In this

chapter we also briefly touch on alternatives to DSP processors, such as fixed-function DSP integrated circuits.

- **DSP Processor Characteristics**
  Chapters three through 17 examine the characteristics of DSP processors in depth. These chapters can be read in their entirety, in which case they provide a comprehensive treatment of modern DSP processor features, or they can be used to explore certain processor characteristics that are important for a particular application.
- **Conclusions**
  In Chapter 18 we present our conclusions on strategies for comparing processors, the current state of the art in DSP processors, and likely future developments in DSP processor technology.
- **Vendor Contact Information**
  The Appendix contains addresses and telephone numbers for companies that sell DSP processors.
- **References and Bibliography, Glossary, and Index**
  The references list many useful sources of information for those interested in delving more deeply into the topics covered here. A glossary of DSP processor-related terms provides definitions of technical terminology used in this book. To help you quickly find the information you need, an extensive index is included at the end of this book.
- **About the Authors**

Acknowledgments

This book would not have been possible without the help of many people.

First, we would like to thank the members and former members of our staff at Berkeley Design Technology. Those who contributed through their efforts in benchmark coding, processor analysis, document production, or work on previous industry reports include Franz Weller, Mohammed Kabir, Rosemary Brock, Dave Wilson, Cynthia Keller, Stephen Slater, and Michael Kiernan. Ivan Heling of TechScribe provided valuable editing and document production services.

Second, we would like to acknowledge the assistance of the many individuals too numerous to list here who contributed helpful insights and advice. In particular, we wish to thank Will Strauss of Forward Concepts for sharing with us his encyclopedic knowledge of the DSP industry. Kennard White provided valuable insights over the course of many wide-ranging discussions about DSP processors, communications, and the DSP industry as a whole.

Finally, we thank product marketing managers, applications engineers, and designers at Analog Devices, AT&T Microelectronics, Clarkspur Design, DSP Group, IBM Microelectronics, Motorola, NEC, SGS-Thomson, TCSI, Texas Instruments, 3Soft, Zilog, and Zoran. All gave generously of their time and expertise.

Chapter 1

Digital Signal Processing and DSP Systems

For the purposes of this book, we define a digital signal processing (DSP) system to be any electronic system making use of digital signal processing. Our informal definition of digital signal processing is the application of mathematical operations to digitally represent signals. Signals are represented digitally as sequences of *samples*. Often, these samples are obtained from physical signals (for example, audio signals) through the use of *transducers* (such as microphones) and *analog-to-digital converters*. After mathematical processing, digital signals may be converted back to physical signals via *digital-to-analog converters*.

In some systems, the use of DSP is central to the operation of the system. For example, modems and digital cellular telephones rely very heavily on DSP technology. In other products, the use of DSP is less central, but often offers important competitive advantages in terms of features, performance, and cost. For example, manufacturers of primarily analog consumer electronics devices like audio amplifiers are beginning to employ DSP technology to provide new features.

This chapter presents a high-level overview of digital signal processing. We first discuss the advantages of DSP over analog systems. We then describe some salient features and characteristics of DSP systems in general. We conclude with a brief look at some important classes of DSP applications.

This chapter is not intended to be a tutorial on DSP theory. For a general introduction to DSP theory, we recommend one of the many textbooks now available on DSP, such as *Discrete-Time Signal Processing* by Oppenheim and Schafer [Opp89].

1.1 Advantages of DSP

Digital signal processing enjoys several advantages over analog signal processing. The most significant of these is that DSP systems are able to accomplish tasks inexpensively that would be difficult or even impossible using analog electronics. Examples of such applications include speech synthesis, speech recognition, and high-speed modems involving error-correction coding. All of these tasks involve a combination of signal processing and control (e.g., making decisions regarding received bits or received speech) that is extremely difficult to implement using analog techniques.

DSP systems also enjoy two additional advantages over analog systems:

- **Insensitivity to environment.** Digital systems, by their very nature, are considerably less sensitive to environmental conditions than analog systems. For example, an analog circuit's behavior depends on its temperature. In contrast, barring catastrophic failures, a DSP system's operation does not depend on its environment—whether in the snow or in the desert, a DSP system delivers the same response.
- **Insensitivity to component tolerances.** Analog components are manufactured to particular tolerances—a resistor, for example, might be guaranteed to have a resistance within 1 percent of its nominal value. The overall response of an analog system depends on the actual values of all of the analog components used. As a result, two analog systems of exactly the same design will have slightly different responses due to slight variations in their components. In contrast, correctly functioning digital components always produce the same outputs given the same inputs.

These two advantages combine synergistically to give DSP systems an additional advantage over analog systems:

- **Predictable, repeatable behavior.** Because a DSP system's output does not vary due to environmental factors or component variations, it is possible to design systems having exact, known responses that do not vary.

Finally, some DSP systems may also have two other advantages over analog systems:

- **Reprogrammability.** If a DSP system is based on programmable processors, it can be reprogrammed—even in the field—to perform other tasks. In contrast, analog systems require physically different components to perform different tasks.
- **Size.** The size of analog components varies with their values; for example, a 100 μF capacitor used in an analog filter is physically larger than a 10 pF capacitor used in a different analog filter. In contrast, DSP implementations of both filters might well be the same size—indeed, might even use the same hardware, differing only in their filter coefficients—and might be smaller than either of the two analog implementations.

These advantages, coupled with the fact that DSP can take advantage of the rapidly increasing density of digital integrated circuit (IC) manufacturing processes, increasingly make DSP the solution of choice for signal processing.

1.2 Characteristics of DSP Systems

In this section we describe a number of characteristics common to all DSP systems, such as algorithms, sample rate, clock rate, and arithmetic types.

Algorithms

DSP systems are often characterized by the *algorithms* used. The algorithm specifies the arithmetic operations to be performed but does not specify how that arithmetic is to be implemented. It might be implemented in software on an ordinary microprocessor or programmable signal processor, or it might be implemented in custom integrated circuits. The selection of an implementation technology is determined in part by the required speed and arithmetic precision. Table 1-1 lists some common types of DSP algorithms and some applications in which they are typically used.

TABLE 1-1. Common DSP Algorithms and Typical Applications

DSP Algorithm	System Application
Speech coding and decoding	Digital cellular telephones, personal communications systems, digital cordless telephones, multimedia computers, secure communications
Speech encryption and decryption	Digital cellular telephones, personal communications systems, digital cordless telephones, secure communications
Speech recognition	Advanced user interfaces, multimedia workstations, robotics, automotive applications, digital cellular telephones, personal communications systems, digital cordless telephones
Speech synthesis	Multimedia PCs, advanced user interfaces, robotics
Speaker identification	Security, multimedia workstations, advanced user interfaces
Hi-fi audio encoding and decoding	Consumer audio, consumer video, digital audio broadcast, professional audio, multimedia computers
Modem algorithms	Digital cellular telephones, personal communications systems, digital cordless telephones, digital audio broadcast, digital signaling on cable TV, multimedia computers, wireless computing, navigation, data/facsimile modems, secure communications
Noise cancellation	Professional audio, advanced vehicular audio, industrial applications
Audio equalization	Consumer audio, professional audio, advanced vehicular audio, music
Ambient acoustics emulation	Consumer audio, professional audio, advanced vehicular audio, music
Audio mixing and editing	Professional audio, music, multimedia computers
Sound synthesis	Professional audio, music, multimedia computers, advanced user interfaces
Vision	Security, multimedia computers, advanced user interfaces, instrumentation, robotics, navigation
Image compression and decompression	Digital photography, digital video, multimedia computers, video-over-voice, consumer video
Image compositing	Multimedia computers, consumer video, advanced user interfaces, navigation
Beamforming	Navigation, medial imaging, radar/sonar, signals intelligence
Echo cancellation	Speakerphones, modems, telephone switches
Spectral estimation	Signals intelligence, radar/sonar, professional audio, music

Sample Rates

A key characteristic of a DSP system is its *sample rate:* the rate at which samples are consumed, processed, or produced. Combined with the complexity of the algorithms, the sample rate determines the required speed of the implementation technology. A familiar example is the digital audio compact disc (CD) player, which produces samples at a rate of 44.1 kHz on two channels.

Of course, a DSP system may use more than one sample rate; such systems are said to be *multirate DSP systems*. An example is a converter from the CD sample rate of 44.1 kHz to the digital audio tape (DAT) rate of 48 kHz. Because of the awkward ratio between these sample rates, the conversion is usually done in stages, typically with at least two intermediate sample rates. Another example of a multirate algorithm is a filter bank, used in applications such as speech, audio, and video encoding and some signal analysis algorithms. Filter banks typically consist of stages that divide the signal into high- and low-frequency portions. These new signals are then downsampled (i.e., their sample rate is lowered by periodically discarding samples) and divided again. In multirate applications, the ratio between the highest and the lowest sample rates in the system can become quite large, sometimes exceeding 100,000.

The range of sample rates encountered in signal processing systems is huge. In Figure 1-1 we show the rough positioning of a few classes of applications with respect to algorithm complexity and sample rate. Sample rates for applications range over 12 orders of magnitude! Only at the very top of that range is digital implementation rare. This is because the cost and difficulty of implementing a given algorithm digitally increases with the sample rate. DSP algorithms used at higher sample rates tend to be simpler than those used at lower sample rates.

Many DSP systems must meet extremely rigorous speed goals, since they operate on lengthy segments of real-world signals in real-time. Where other kinds of systems (like databases) may be required to meet performance goals *on average*, real-time DSP systems often must meet such goals *in every instance*. In such systems, failure to maintain the necessary processing rates is considered a serious malfunction. Such systems are often said to be subject to *hard real-time constraints*. For example, let's suppose that the compact disc-to-digital audio tape sample rate converter discussed above is to be implemented as a real-time system, accepting digital signals at the CD sample rate of 44.1 kHz and producing digital signals at the DAT sample rate of 48 kHz. The converter must be ready to accept a new sample from the CD source *every* 22.6 μs (i.e., 1/44100 s), and must produce a new output sample for the DAT device *every* 20.8 μs (1/48000 s). If the system ever fails to accept or produce a sample on this schedule, data are lost and the resulting output signal is corrupted. The need to meet such hard real-time constraints creates special challenges in the design and debugging of real-time DSP systems.

Clock Rates

Digital electronic systems are often characterized by their *clock rates*. The clock rate usually refers to the rate at which the system performs its most basic unit of work. In mass-produced, commercial products, clock rates of up to 100 MHz are common, with faster rates found in some high-performance products. For DSP systems, the ratio of system clock rate to sample rate is one of the most important characteristics used to determine how the system will be implemented. The relationship between the clock rate and the sample rate partially determines the amount of hard-

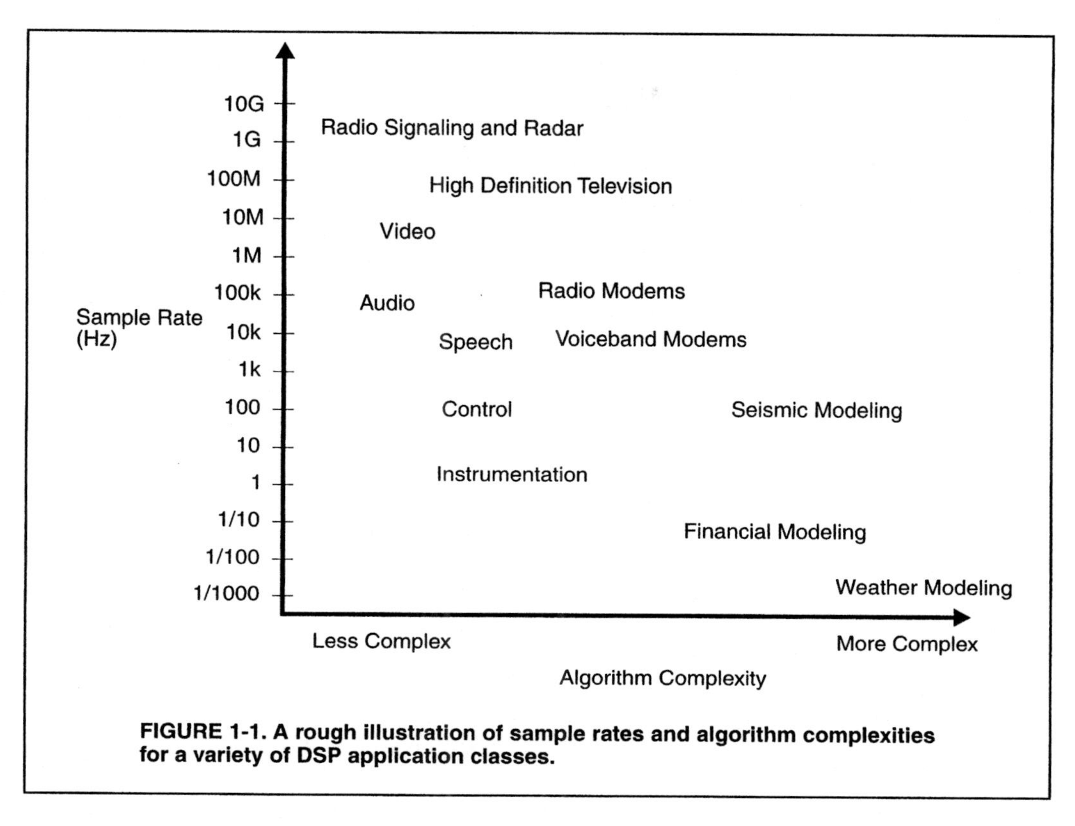

FIGURE 1-1. A rough illustration of sample rates and algorithm complexities for a variety of DSP application classes.

ware needed to implement an algorithm with a given complexity in real-time. As the ratio of sample rate to clock rate increases, so does the amount and complexity of hardware required to implement the algorithm.

Numeric Representations

Arithmetic operations such as addition and multiplication are at the heart of DSP algorithms and systems. As a result, the numeric representations and type of arithmetic used can have a profound influence on the behavior and performance of a DSP system. The most important choice for the designer is between fixed-point and floating-point arithmetic. Fixed-point arithmetic represents numbers in a fixed range (e.g., –1.0 to +1.0) with a finite number of bits of precision (called the *word width*). For example, an eight-bit fixed-point number provides a resolution of 1/256 of the range over which the number is allowed to vary. Numbers outside of the specified range cannot be represented; arithmetic operations that would result in a number outside this range either *saturate* (that is, are limited to the largest positive or negative representable value) or *wrap around* (that is, the extra bits resulting from the arithmetic operation are discarded).

Floating-point arithmetic greatly expands the representable range of values. Floating-point arithmetic represents every number in two parts: a mantissa and an exponent. The mantissa is, in effect, forced to lie between –1.0 and +1.0, while the exponent keeps track of the amount by which the mantissa must be scaled (in terms of powers of two) in order to create the actual value represented. That is:

$$value = mantissa \times 2^{exponent}$$

Floating-point arithmetic provides much greater dynamic range (that is, the ratio between the largest and smallest values that can be represented) than fixed-point arithmetic. Because it reduces the probability of overflow and the necessity of scaling, it can considerably simplify algorithm and software design. Unfortunately, floating-point arithmetic is generally slower and more expensive than fixed-point arithmetic, and is more complicated to implement in hardware than fixed-point arithmetic.

Arithmetic and numeric formats are discussed in more detail in Chapter 3.

1.3 Classes of DSP Applications

Digital signal processing in general, and DSP processors in particular, are used in an extremely diverse range of applications, from radar systems to consumer electronics. Naturally, no one processor can meet the needs of all or even most applications. Therefore, the first task for the designer selecting a DSP processor is to weigh the relative importance of performance, cost, integration, ease of development, power consumption, and other factors for the application at hand. Here we briefly touch on the needs of just a few categories of DSP applications. Table 1-2 summarizes these categories.

Low-Cost Embedded Systems

The largest applications (in terms of dollar volume) for digital signal processors are inexpensive, high-volume embedded systems, such as cellular telephones, disk drives (where DSPs are used for servo control), and modems. In these applications, cost and integration considerations are paramount. For portable, battery-powered products, power consumption is also critical. In these high-volume, embedded applications, performance and ease of development considerations

TABLE 1-2. Example DSP Processor Applications

Category	Example Applications
Low-cost embedded systems	Modems, radar detectors, pagers, cellular telephones, cordless telephones, disk drives, automotive real-time control
High-performance applications	Radar, sonar, seismic imaging, speaker identification
Personal computer-based multimedia	Modems, voice mail, music synthesis, speech synthesis, speech and audio compression and decompression

are often given less weight, even though these applications usually involve the development of custom software to run on the DSP and custom hardware that interfaces with the DSP.

High-Performance Applications

Another important class of applications involves processing large volumes of data with complex algorithms for specialized needs. This includes uses like sonar and seismic exploration, where production volumes are lower, algorithms are more demanding, and product designs are larger and more complex. As a result, designers favor processors with maximum performance, ease of use, and support for multiprocessor configurations. In some cases, rather than designing their own hardware and software from scratch, designers of these systems assemble systems using standard development boards and ease their software development tasks by using existing software libraries.

Personal Computer-Based Multimedia

A newer class of applications is personal computer-based multimedia functions. Increasingly, PCs are incorporating DSP processors to provide such varied capabilities as voice mail, data and facsimile modems, music and speech synthesis, and image compression. As with other high-volume, embedded applications, PC multimedia demands low cost and high integration. Unlike some other embedded applications, PC multimedia also demands high performance, since a DSP processor in a multimedia PC may be called on to perform multiple functions simultaneously. In addition, the multitasking nature of such applications means that, in addition to performing each function efficiently, the DSP processor must have the ability to efficiently switch between functions. Memory capacity may also be an issue in these applications, because many multimedia applications require the ability to manipulate large amounts of data.

An increasing trend is the incorporation of DSP-like functions into general-purpose processors to better handle signal processing tasks. In some cases, such augmented microprocessors may be able to handle certain tasks without the need of a separate DSP processor. However, we believe the use of separate general-purpose processors and DSP processors has much to offer and will continue to be the dominant implementation of PC-based multimedia for some time to come.

Chapter 2

DSP Processors, Embodiments, and Alternatives

The previous chapter described digital signal processing in general terms, focusing on DSP fundamentals, systems, and application areas. In this chapter we narrow our focus to DSP processors. We begin with a high-level description of the features common to virtually all DSP processors. We then describe typical embodiments of DSP processors and briefly discuss alternatives to DSP processors, such as general-purpose microprocessors. The next several chapters provide a detailed treatment of DSP processor architectures and features.

2.1 DSP Processors

Dozens of families of DSP processors are available on the market today. The salient features of some of the more notable families of DSP processors are summarized in Table 2-1. Throughout this book, these processors are used as examples to illustrate the architectures and features that can be found in commercial DSP processors.

Most DSP processors share some common features designed to support repetitive, numerically intensive tasks. The most important of these features are introduced briefly here and summarized in Table 2-2. Each of these features and many others are examined in greater detail in the later chapters of this book.

Fast Multiply-Accumulate

The most often cited feature of DSP processors is the ability to perform a *multiply-accumulate* operation (often called a *MAC*) in a single instruction cycle. The multiply-accumulate operation is useful in algorithms that involve computing a vector product, such as digital filters, correlation, and Fourier transforms. To achieve this functionality, DSP processors include a multiplier and accumulator integrated into the main arithmetic processing unit (called the *data path*) of the processor. In addition, to allow a series of multiply-accumulate operations to proceed without the possibility of arithmetic overflow, DSP processors generally provide extra bits in their accumulator registers to accommodate growth of the accumulated result. DSP processor data paths are discussed in detail in Chapter 4, "Data Path."

TABLE 2-1. Commercial DSP Processors (Excluding DSP Cores) Used as Examples in This Book

Vendor	Processor Family	Arith. Type	Data Width	Speed (MIPS)
Analog Devices	ADSP-21xx	Fixed	16	33.3
	ADSP-210xx	Floating	32	40.0
AT&T	DSP16xx	Fixed	16	70.0
	DSP32xx	Floating	32	20.0
Motorola	DSP5600x	Fixed	24	40.0
	DSP561xx	Fixed	16	30.0
	DSP563xx	Fixed	24	80.0
	DSP96002	Floating	32	20.0
NEC	µPD7701x	Fixed	16	33.3
Texas Instruments	TMS320C1x	Fixed	16	8.8
	TMS320C2x	Fixed	16	12.5
	TMS320C2xx	Fixed	16	40.0
	TMS320C3x	Floating	32	25.0
	TMS320C4x	Floating	32	30.0
	TMS320C5x	Fixed	16	50.0
	TMS320C54x	Fixed	16	50.0
	TMS320C8x	Fixed	8/16	50.0
Zoran	ZR3800x	Fixed	20	33.3

Multiple-Access Memory Architecture

A second feature shared by most DSP processors is the ability to complete several accesses to memory in a single instruction cycle. This allows the processor to fetch an instruction while simultaneously fetching operands for the instruction or storing the result of the previous instruction to memory. High bandwidth between the processor and memory is essential for good performance if repetitive data-intensive operations are required in an algorithm, as is common in many DSP applications.

TABLE 2-2. Basic Features Common to Virtually all DSP Processors

Feature	Use
Fast multiply-accumulate	Most DSP algorithms, including filtering, transforms, etc. are multiplication-intensive.
Multiple-access memory architecture	Many data-intensive DSP operations require reading a program instruction and multiple data items during each instruction cycle for best performance.
Specialized addressing modes	Efficient handling of data arrays and first-in, first-out buffers in memory.
Specialized program control	Efficient control of loops for many iterative DSP algorithms. Fast interrupt handling for frequent I/O operations.
On-chip peripherals and input/output interfaces	On-chip peripherals like analog-to-digital converters allow for small, low-cost system designs. Similarly, I/O interfaces tailored for common peripherals allow clean interfaces to off-chip I/O devices.

In many processors, single-cycle multiple memory accesses are subject to restrictions. Typically, all but one of the memory locations accessed must reside on-chip, and multiple memory accesses can take place only with certain instructions. To support simultaneous access of multiple memory locations, DSP processors provide multiple on-chip buses, multiported on-chip memories, and in some cases, multiple independent memory banks. DSP memory structures are quite distinct from those of general-purpose processors. DSP processor memory architectures are investigated in detail in Chapter 5, "Memory Architecture."

Specialized Addressing Modes

To allow arithmetic processing to proceed at maximum speed and to allow specification of multiple operands in a small instruction word, DSP processors incorporate dedicated address generation units. Once the appropriate addressing registers have been configured, the address generation units operate in the background, forming the addresses required for operand accesses in parallel with the execution of arithmetic instructions. Address generation units typically support a selection of addressing modes tailored to DSP applications. The most common of these is register-indirect addressing with post-increment, which is used in situations where a repetitive computation is performed on a series of data stored sequentially in memory. Special addressing modes (called *circular* or *modulo* addressing) are often supported to simplify the use of data buffers. Some processors support bit-reversed addressing, which eases the task of interpreting the results of the fast Fourier transform (FFT) algorithm. Addressing modes are described in more detail in Chapter 6, "Addressing."

Specialized Execution Control

Because many DSP algorithms involve performing repetitive computations, most DSP processors provide special support for efficient looping. Often, a special loop or repeat instruction is provided that allows the programmer to implement a *for-next* loop without expending any instruction cycles for updating and testing the loop counter or for jumping back to the top of the loop.

Some DSP processors provide other execution control features to improve performance, such as fast context switching and low-latency, low-overhead interrupts for fast input/output.

Hardware looping and interrupts are discussed in Chapter 8, "Execution Control."

Peripherals and Input/Output Interfaces

To allow low-cost, high-performance input and output (I/O), most DSP processors incorporate one or more serial or parallel I/O interfaces, and specialized I/O handling mechanisms such as direct memory access (DMA). DSP processor peripheral interfaces are often designed to interface directly with common peripheral devices like analog-to-digital and digital-to-analog converters.

As integrated circuit manufacturing techniques have improved in terms of density and flexibility, DSP processor vendors have begun to include not just peripheral interfaces, but complete peripheral devices on-chip. Examples of this are chips designed for cellular telephone applications, several of which incorporate a digital-to-analog and analog-to-digital converter on-chip.

Various features of DSP processor peripherals are described in Chapter 10, "Peripherals."

2.2 DSP Processor Embodiments

The most familiar form of DSP processor is the single-chip processor, which is incorporated into a printed circuit board design by the system designer. However, with the widespread proliferation of DSP processors into many new kinds of applications, the increasing levels of integration in all kinds of electronic products, and the development of new packaging techniques, DSP processors can now be found in many different forms, sometimes masquerading as something else entirely. In this section we briefly discuss some of the forms that DSP processors take.

Multichip Modules

Multichip modules (MCMs) are a kind of superchip. Rather than packaging a single integrated circuit (IC) die in a ceramic or plastic package as is done with conventional ICs, MCMs combine multiple, bare (i.e., unpackaged) dies into a single package. One advantage of this approach is achieving higher packaging density—more circuits per square inch of printed circuit board. This, in turn, results in increased operating speed and reduced power dissipation. As multichip module packaging technology has advanced in the past few years, vendors have begun to offer multichip modules containing DSP processors. For example, Texas Instruments sells an MCM that includes two TMS320C40 processors and 128 Kwords of 32-bit static random-access memory (SRAM).

Multiple Processors on a Chip

As integrated circuit manufacturing technology becomes more sophisticated, DSP processor designers can squeeze more features and performance onto a single-chip processor, and they can consider combining multiple processors on a single integrated circuit. For example, both Motorola and Zilog offer parts that combine a DSP and microprocessor or microcontroller on a single chip, a natural combination for many applications. As with multichip modules, multiprocessor chips provide increased performance and reduced power compared with designs using multiple, separately packaged processors. However, the selection of multiprocessor chip offerings is currently limited to only a few devices.

Chip Sets

While some manufacturers combine multiple processors on a single chip, and others use multichip modules to combine multiple chips into one package, another variation on DSP processor packaging is to divide the DSP into two or more separate packages. This is the approach that Butterfly DSP has taken with their DSP chip set, which consists of the LH9320 address generator and the LH9124 processor (formerly sold by Sharp Microelectronics). Dividing the processor into two or more packages may make sense if the processor is very complex and if the number of input/output pins is very large. Splitting functionality into multiple integrated circuits may allow the use of much less expensive IC packages, and thereby provide cost savings. This approach also provides added flexibility, allowing the system designer to combine the individual ICs in the configuration best suited for the application. For example, with the Butterfly chip set, multiple address generator chips can be used in conjunction with one processor chip. Finally, chip sets have the potential of providing more I/O pins than individual chips. In the case of the Butterfly chip set, the use of separate address generator and processor chips allows the processor to have eight 24-bit external data buses, many more than provided by more common single-chip processors.

DSP Cores

An interesting approach for high-volume designs is the coupling of a programmable DSP with user-defined circuitry on a single chip. This approach combines the benefits of a DSP processor (such as programmability, development tools, and software libraries) with the benefits of custom circuits (e.g., low production costs, small size, and low power consumption). In this section, we briefly describe two variants of this design style: DSP core-based application-specific integrated circuits (ASICs) and customizable DSP processors.

A DSP core is a DSP processor intended for use as a building block in creating a chip, as opposed to being packaged by itself as an off-the-shelf chip. A DSP core-based ASIC is an ASIC that incorporates a DSP core as one element of the overall chip. The DSP core-based ASIC approach allows the system designer to integrate a programmable DSP, interface logic, peripherals, memory, and other custom elements onto a single integrated circuit. Figures 2-1 and 2-2 illustrate the DSP core-based ASIC concept.

Many vendors of DSP processors use the core-based approach to create versions of their standard processors targeted at entire application areas, like telecommunications. In our discus-

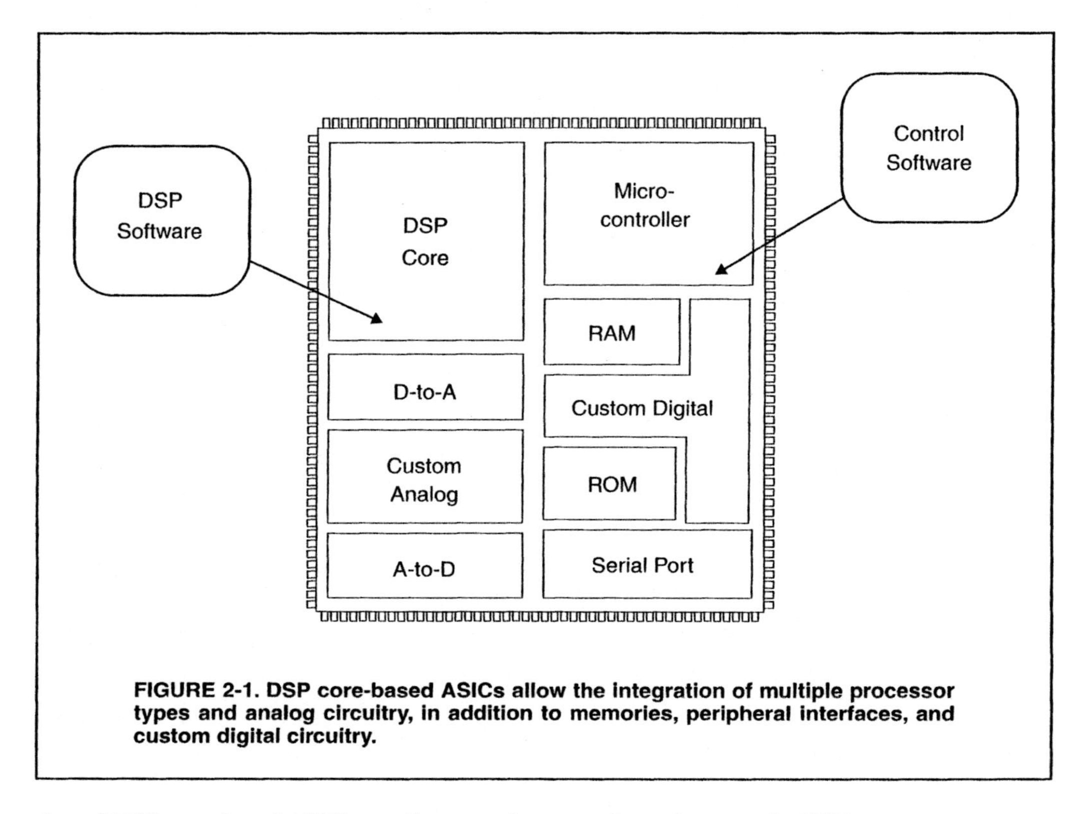

FIGURE 2-1. DSP core-based ASICs allow the integration of multiple processor types and analog circuitry, in addition to memories, peripheral interfaces, and custom digital circuitry.

sion of DSP core-based ASICs, we focus on the case where the user of a DSP processor wants to create a DSP core-based ASIC for a specific application.

There are presently several companies established as providers of DSP cores for use in ASIC designs: AT&T, Texas Instruments, SGS-Thomson Microelectronics, DSP Group, Clarkspur Design, 3Soft, and Tensleep Design. Two additional firms (TCSI and Infinite Solutions) have announced plans to offer DSP cores as well. Currently available DSP cores are summarized in Table 2-3.

Note that vendors differ in their definitions of exactly what is included in a "DSP core." For example, Texas Instruments' definition of a DSP core includes not only the processor, but memory and peripherals as well. Clarkspur Design's and DSP Group's cores include memory but not peripherals. SGS-Thomson's core includes only the processor and no peripherals or memory.

The services that these companies provide can be broadly divided into two categories. In the first category, the vendor of the core is also the provider of foundry services used to fabricate the ASIC containing the core; we refer to this category as "foundry-captive." In the second category, the core vendor licenses the core design to the customer, who is then responsible for selecting an appropriate foundry. We call this category "licensable."

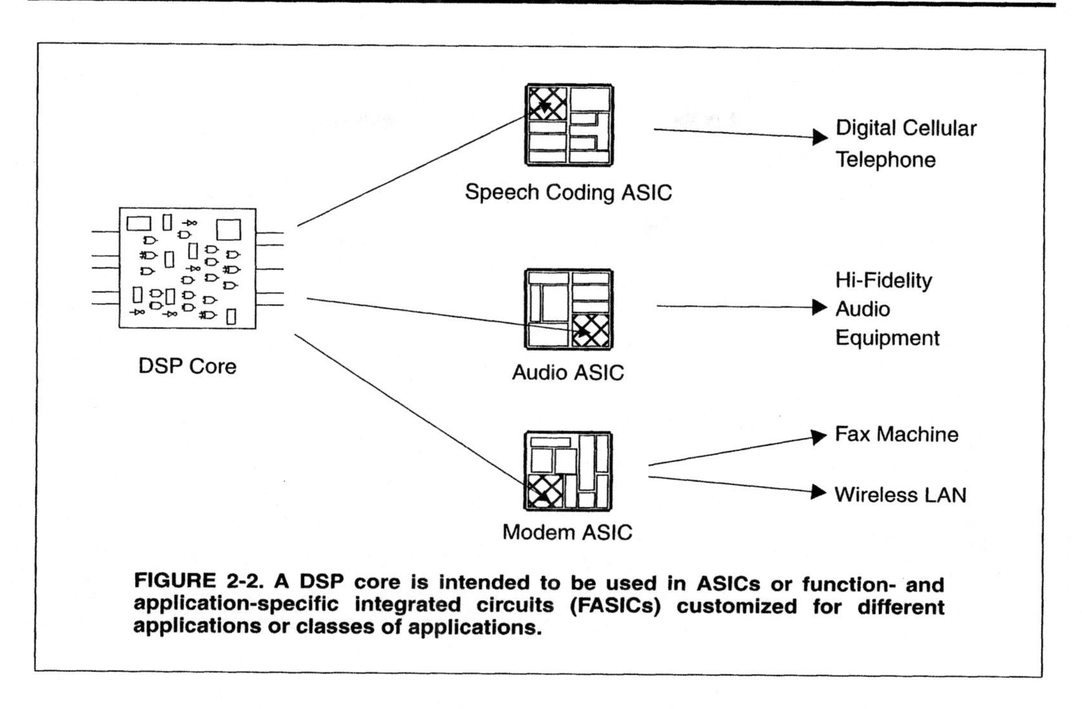

FIGURE 2-2. A DSP core is intended to be used in ASICs or function- and application-specific integrated circuits (FASICs) customized for different applications or classes of applications.

TABLE 2-3. Currently Available DSP Cores. *

Vendor	Core Family	Arith. Type	Data Width	Speed (MIPS)
Clarkspur Design	CD2400	Fixed	16	25.0
	CD2450	Fixed	16-24	50.0
DSP Group	PineDSPCore	Fixed	16	40.0
	OakDSPCore	Fixed	16	40.0
SGS-Thomson	D950-CORE	Fixed	16	40.0
Tensleep Design	A/DSCx21	Fixed	16	30.0
Texas Instruments	T320C2xLP	Fixed	16	40.0
	TEC320C52	Fixed	16	50.0
3Soft	M320C25	Fixed	16	15.0

* Several companies, including Infinite Solutions and TCSI, have announced plans to offer DSP cores that are not listed here.

Foundry-Captive DSP Cores

Texas Instruments and SGS-Thomson are both providers of foundry-captive DSP cores. Texas Instruments has several design approaches for DSP core-based ASICs. For lower volume designs, Texas Instruments offers the TEC320C52, which surrounds its TMS320C52 16-bit fixed-point DSP with a gate array containing 15,000 gates. The customer designs custom circuitry to surround the core, which is then implemented (typically using logic synthesis tools) in the gate array portion of the chip. Texas Instruments then fabricates the chip.

The "DSP core embedded within a gate array" approach enjoys advantages in the areas of design cost, production cost, and manufacturing lead time. Because of the reduced complexity of the design process for gate arrays, it can proceed faster than the design process for other kinds of ICs. And because the only custom aspect of the integrated circuit is the metal layers, which are added at the end of the fabrication process, larger economies of scale can result in reduced cost. Finally, die prefabrication results in faster manufacturing of finished parts.

For higher volume designs, Texas Instruments also supports full- or semi-custom chip designs using their standard DSP processors (e.g., TMS320C1x, TMS320C2x, TMS320C2xx, TMS320C3x, and TMS320C5x) as cores (macrocells) within a chip. These cores can be surrounded by full-custom layouts, standard cells, gate arrays, or a mixture of these design options.

SGS-Thomson offers their D950-CORE (a 16-bit fixed-point DSP core) as a macrocell in their standard ASIC library. Application-specific hardware designed by the customer is crafted from standard cells or in a gate array.

Licensable DSP Cores

In contrast to Texas Instruments and SGS-Thomson Microelectronics, companies such as DSP Group, Clarkspur Design, Infinite Solutions, Tensleep Design, TCSI, and 3Soft offer *licensable* cores. In exchange for a license fee and (in some cases) royalties, the customer receives a complete design description of the DSP core. This core can then be fabricated as part of an ASIC using the IC foundry of the customer's choice. Additionally, the customer can modify the core processor if desired, since the complete design is available.

Licensable cores are generally provided in the form of an optimized full-custom layout compatible with the fabrication processes of a particular foundry. Synthesizable VHDL or Verilog HDL design descriptions are often available from the core vendor as well.

Customizable DSP Processors

In some cases, it may be desirable to modify or extend the DSP core itself (as opposed to adding external circuitry surrounding the core). For example, a designer might want to add a new functional unit (such as an error correction coding unit or bit manipulation unit) to the data path of the core to improve the core's performance in an application. We term a DSP processor that supports such modifications a *customizable DSP processor.*[1]

[1] Our use of the term "customizable DSP processor" should not be confused with Texas Instruments' "customizable DSP" (cDSP), which is their term for DSP core-based ASICs using Texas Instruments DSP cores.

As mentioned earlier, different vendors use the term "core" in different ways, so a "customizable DSP core" may have a range of meanings, from simply allowing the addition of peripherals and memory to support for addition to or modification of the processor's execution units.

Although customizing a DSP core has a potentially large payoff, it also poses serious drawbacks. First, in the case of a foundry-captive core, the customer may not have access to the internal design of the core. As a result, the desired modifications must be made by the chip vendor—a potentially expensive proposition. In the case of a licensable core, the design is accessible to the customer and can be modified by the customer's engineers. However, these engineers may not be sufficiently expert in the core's internal design to efficiently make the desired modifications. Finally, changes to the core processor architecture require that corresponding changes be made to the core's software development tools (assembler, linker, simulator, and so on).

At present, the companies that have best addressed the challenges of customizable DSP processors are AT&T Microelectronics and Philips—neither of which offers their cores in a broad fashion. In particular, AT&T's DSP1600 processor core was designed to permit easy attachment of new execution units to its data path, and its software development tools were designed to facilitate the addition of support for new execution units into the tools. Similarly, Philips' EPICS core was designed with configurability in mind; Philips has demonstrated versions of the core with different word widths.

Multiprocessors

No matter how fast and powerful DSP processors become, the needs of a large class of important applications cannot be met by a single processor. Some of these applications may be well suited to custom integrated circuits. If programmability is important, a multiprocessor based on commercial DSPs may be an effective solution. Although any DSP processor can be used in a multiprocessor design, some manufacturers have made special efforts to create DSPs that are especially well suited to multiprocessor systems. Among these are Texas Instruments' TMS320C4x, Motorola's DSP96002, and Analog Devices' ADSP-2106x. These processors include features such as multiple external buses, bus-sharing logic, and (on the Texas Instruments and Analog Devices processors) multiple, dedicated parallel ports designed for interprocessor communication that simplify hardware design and improve performance.

2.3 Alternatives to Commercial DSP Processors

There are many alternatives to DSP processors available to the designer contemplating a new application. Although we do not analyze these alternatives in detail in this book, a few of them are outlined briefly below. Many of these approaches are explored in depth in the report, *DSP Design Tools and Methodologies* [BDT95], also from Berkeley Design Technology.

General-Purpose Microprocessors in Embedded Systems

As general-purpose microprocessors become faster, they are increasingly able to handle some less-demanding DSP applications. Although general-purpose processors are much slower and less efficient than DSP processors for DSP applications, there may still be strong reasons to use a general-purpose processor for a given application. First, many electronics products—from telephones to automotive engine controllers—are currently designed using general-purpose microprocessors for control, user interface, and communications functions. If a DSP application is being added to an existing product that already contains a general-purpose microprocessor, then it may be possible to add the new application without needing an additional processor. This approach has obvious cost advantages, though it is only suited to relatively simple DSP applications. Second, software development tools for general-purpose processors are generally much more sophisticated and powerful than those for DSP processors. For applications that are relatively undemanding in terms of performance, but for which ease of development is a critical consideration, this can be an important factor.

As digital signal processing-intensive applications increasingly move into the mainstream of computing and electronics products, general-purpose processors have begun to adopt some of the features of DSP processors to make them more suitable for DSP-intensive applications. For example, the Motorola/IBM PowerPC 601, the MIPS R10000, the Sun UltraSPARC, and the Hewlett-Packard PA-7100LC general-purpose microprocessors are all able to perform a floating-point multiply-accumulate in one instruction cycle under certain circumstances. Additionally, some of these processors have special instructions aimed at multimedia signal processing applications. Similarly, Intel has announced a version of the Pentium with features designed to better support DSP.

It isn't yet clear, though, whether a single, hybrid processor or separate general-purpose processor and DSP processor (possibly on the same chip) will become the more popular approach. We expect that for the next few years, at least, applications with a significant real-time DSP content will be better served by a separate, specialized DSP processor.

Personal Computers and Workstations

DSP applications may take advantage of entire platforms built upon general-purpose microprocessors. Personal computers and workstations, which are among the most visible of these platforms, introduce special considerations for applications incorporating signal processing.

Some DSP applications may be directly implemented on PCs or workstations that are not equipped with DSP processors. They encounter many of the same performance and efficiency shortcomings mentioned in the preceding section. But, in some cases, the platform with its general-purpose processor is adequate by itself to implement the complete application. Software-only DSP-based products are appropriate where the signal processing requirements are not demanding and where a PC or workstation is already available to the end user. Many scientific and engineering applications that involve non-real-time synthesis or analysis of signals take this approach and run on conventional PCs and workstations.

Perhaps the most visible example of this approach to signal processing is Intel's *native signal processing* (NSP) initiative. NSP seeks to use the host ("native") processor in PC-compatible computers for low-end multimedia applications, such as audio compression and decompression, music and sound synthesis, and more. The obvious advantage of this approach is that it reduces or eliminates the cost and burden associated with purchasing the extra hardware for the PC (such as sound cards) that would otherwise be required.

There is a trend among both PC and workstation manufacturers to add more DSP capabilities to their products, both through on-board DSP processors and through the addition of peripherals like analog-to-digital and digital-to-analog converters and telephone line interfaces that support DSP applications like modems, speech recognition, and music synthesis. As these kinds of resources become increasingly prevalent in PCs and workstations, more opportunities will open up for software-only DSP products.

Custom Hardware

There are two important reasons why custom-developed hardware is sometimes a better choice than a DSP processor-based implementation: performance and production cost. In virtually any application, custom hardware can be designed which provides better performance than a programmable processor. Just as DSP processors are more cost-effective for DSP applications than general-purpose processors because of their specialization, custom hardware has the potential to be even more cost-effective due to its more specialized nature. In applications with high sampling rates (for example, higher than 1/100th of the system clock rate), custom hardware may be the only reasonable approach.

For high-volume products, custom hardware may also be less expensive than a DSP processor. This is because a custom implementation places in hardware only those functions needed by the application, whereas a DSP processor requires every application to pay for the full functionality of the processor, even if it uses only a small subset of its capabilities. Of course, developing custom hardware has some serious drawbacks in addition to these advantages. Most notable among these drawbacks are the effort and expense associated with custom hardware development, especially for custom chip design.

Custom hardware can take many forms. It can be a simple, small printed circuit board using off-the-shelf components, or it can be a complex, multiboard system, incorporating custom integrated circuits. The aggressiveness of the design approach depends on the needs of the application. For an in-depth exploration of DSP system design alternatives and tools, see *DSP Design Tools and Methodologies* [BDT95]. In the remainder of this section we very briefly mention some of the more popular approaches.

One of the most common approaches for custom hardware for DSP applications is to design custom printed circuit boards that incorporate a variety of off-the-shelf components. These components may include standard logic devices, fixed-function or configurable arithmetic units, field-programmable gate arrays (FPGAs), and function- or application-specific integrated circuits (FASICs). As their name implies, FASICs are chips that are designed to perform a specific function, perhaps for a single application. Examples of FASICs include configurable digital filter chips, which can be configured to work in a range of applications, and facsimile modem chips, which are designed specifically to provide the signal processing functions for a fax modem and aren't useful for anything else.

Many off-the-shelf application-specific ICs sold by semiconductor vendors for DSP applications are really standard DSP processors containing custom, mask-programmed software in ROM. Some of these chips are based on commercial DSP processors. For example, Phylon's modem chips are based on Analog Devices and Texas Instruments processors. Others are based on proprietary processor architectures; the most prominent examples of this approach are Rockwell's data and fax modem chips.

As tools for creating custom chips improve and more engineers become familiar with chip design techniques, more companies are developing custom chips for their applications. Designing a custom chip provides the ultimate flexibility, since the chip can be tailored to the needs of the application, down to the level of a single logic gate.

Of course, the benefits of custom chips and other hardware-based implementation approaches come with important trade-offs. Perhaps most importantly, the complexity and cost of developing custom hardware can be high, and the time required can be long. In addition, if the hardware includes a custom programmable processor, new software development tools will be required.

It is important to point out that the implementation options discussed here are not mutually exclusive. In fact, it is quite common to combine many of these design approaches in a single system, choosing different techniques for different parts of the system. One such hybrid approach, DSP core-based ASICs, was mentioned above. Others, such as the combination of an off-the-shelf DSP processor with custom ICs, FPGAs, and a general-purpose processor, are very common.

Chapter 3

Numeric Representations and Arithmetic

One of the most important characteristics determining the suitability of a DSP processor for a given application is the type of binary numeric representation(s) used by the processor. The data representation variations common in commercial DSP processors can be illustrated hierarchically, as in Figure 3-1.

3.1 Fixed-Point versus Floating-Point

The earliest DSP processors used *fixed-point* arithmetic, and in fact fixed-point DSPs still dominate today. In fixed-point processors, numbers are represented either as integers (integer

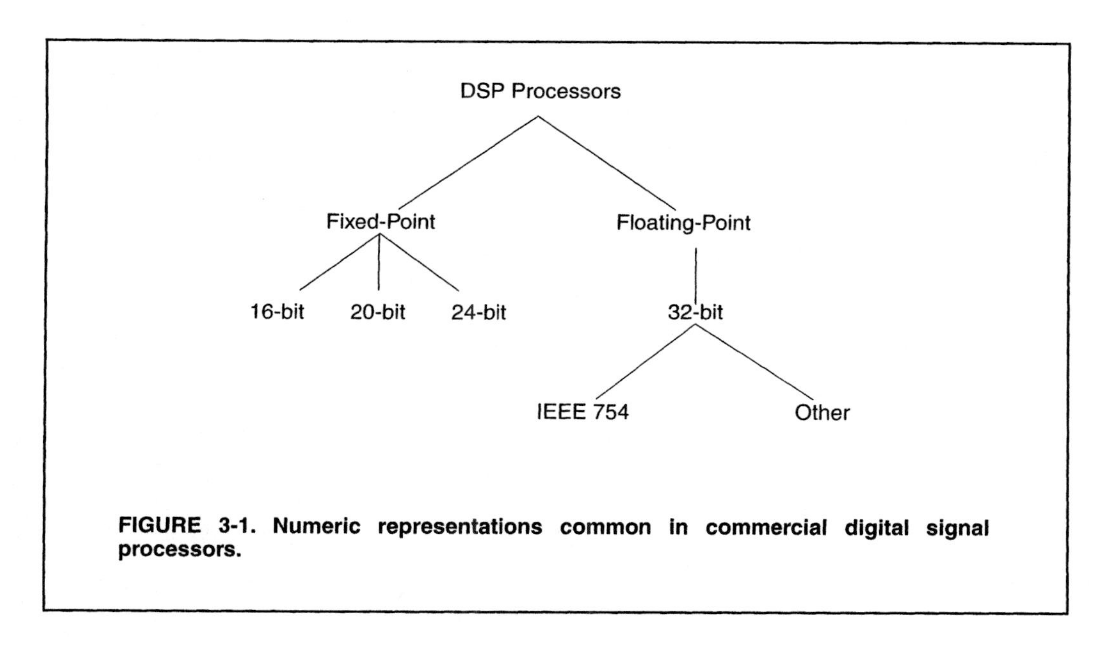

FIGURE 3-1. Numeric representations common in commercial digital signal processors.

arithmetic) or as fractions between –1.0 and +1.0 (fractional arithmetic). The algorithms and hardware used to implement fractional arithmetic are virtually identical to those used for integer arithmetic. The main difference between integer and fractional arithmetic has to do with how the results of multiplication operations are handled. In practice, most fixed-point DSP processors support fractional arithmetic *and* integer arithmetic. The former is most useful for signal processing algorithms, while the latter is useful for control operations, address calculations, and other operations that do not involve signals. Figures 3-2 and 3-3 illustrate simple integer and fractional representations. Note that performing integer arithmetic on fixed-point DSPs sometimes is more time-consuming or more constrained than performing fractional arithmetic.

Another class of DSP processors primarily uses *floating-point* arithmetic, where numbers are represented by the combination of a *mantissa* and an *exponent*. This is illustrated in Figure 3-4. The mantissa is usually a signed fractional value with a single implied integer bit. (The implied integer bit is not actually stored as part of the data value; rather, it is assumed to always be set to "1.") This means that the mantissa can take on a value in the ranges of +1.0 to +2.0 and –1.0 to –2.0. The exponent is an integer that represents the number of places that the binary point of the mantissa (analogous to the decimal point in an ordinary base 10 number) must be shifted left or right to obtain the original number represented. The value represented is computed via an expression of the form:

$$value = mantissa \times 2^{exponent}$$

Figure 3-4 illustrates a simple floating-point data representation. In general, floating-point processors also support fixed-point (usually integer) data formats. This is necessary to facilitate operations that are inherently integer in nature, such as memory address computations.

Floating-point arithmetic is a more flexible and general mechanism than fixed-point. With floating-point, system designers have access to wider *dynamic range* (the ratio between the largest and smallest numbers that can be represented) and in many cases better *precision*.

Our definition of precision is based on the idea of *quantization error*. Quantization error is the numerical error introduced when a longer numeric format is converted to a shorter one. For example, when we round the value 1.325 to 1.33, we have introduced a quantization error of 0.005. The greater the possible quantization error relative to the size of the value represented, the less precision is available.

For a fixed-point format, we define the maximum available precision to be equal to the number of bits in the format. For example, a 16-bit fractional format provides a maximum 16 bits of precision. This definition is based on computing the ratio of the size of the value represented to the size of the maximum quantization error that could be suffered when converting from a more precise representation via rounding. Formally stated,

$$maximum\ precision\ (in\ bits) = log_2\,(|maximum\ value|\ /\ |maximum\ quantization\ error|)$$

For a 16-bit fractional representation, the largest-magnitude value that can be represented is –1.0. When converting to a 16-bit fractional format from a more precise format via rounding, the maximum quantization error is 2^{-16}. Using the relation above, we can compute that this format has a maximum precision of $log_2(1/2^{-16})$, or 16 bits, the same as the format's overall width. Note that if the value being represented has a smaller magnitude than the maximum, the precision obtained is less than the maximum available precision. This underscores the importance of careful signal scaling when using fixed-point arithmetic. Scaling is used to maintain precision by modifying the range of signal values to be near the maximum range of the numeric representation used. Scaling is discussed in detail in Chapter 4, "Data Path."

Using this same definition for a floating-point format, the maximum available precision is the number of bits in the mantissa, including the implied integer bit. Because floating-point processors automatically scale all values so that the implied integer bit is equal to 1, the magnitude of

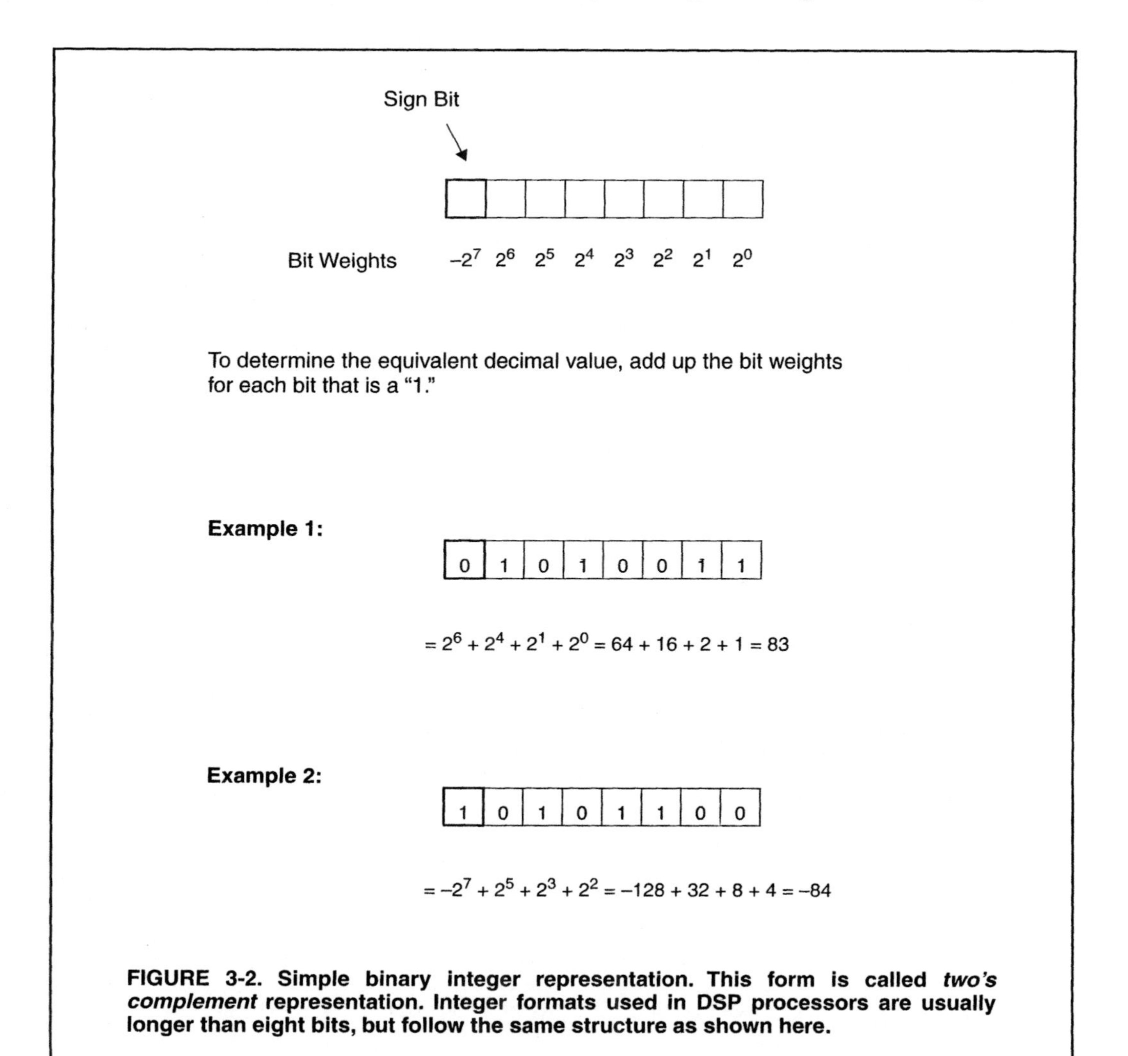

FIGURE 3-2. Simple binary integer representation. This form is called *two's complement* representation. Integer formats used in DSP processors are usually longer than eight bits, but follow the same structure as shown here.

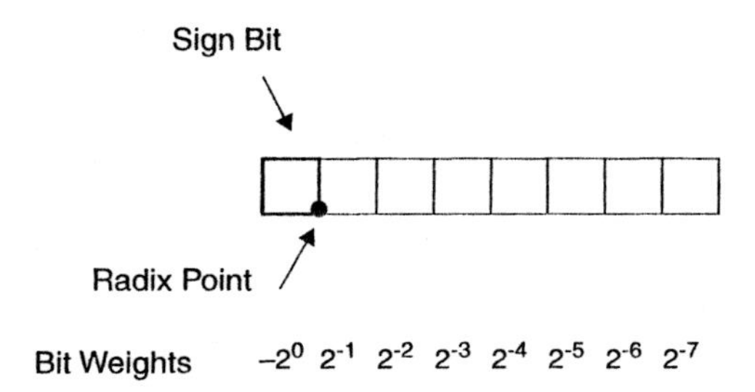

To determine the equivalent decimal value, add up the bit weights for each bit that is a "1."

Example 1:

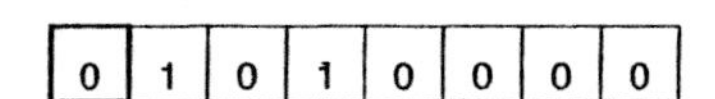

$= 2^{-1} + 2^{-3} = 0.5 + 0.125 = 0.625$

Example 2:

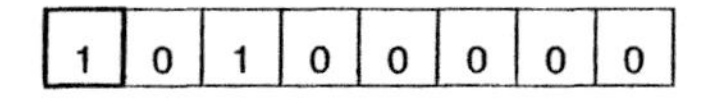

$= -2^0 + 2^{-2} + 2^{-4} = -1.0 + 0.25 + 0.0625 = -0.6875$

FIGURE 3-3. Simple binary fractional representation. This format is identical to the integer format, except that a *radix point* is assumed to exist immediately after the sign bit. Fractional formats used in DSP processors are usually longer than eight bits, but follow the same structure as shown here.

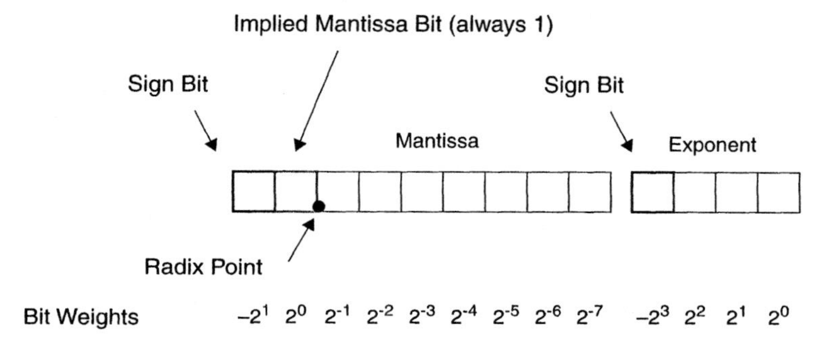

To determine the equivalent decimal value, first compute the mantissa value by adding up the bit weights for the mantissa bits that are "1."

Then, compute the exponent value in the same way.

Finally, multiply the mantissa value by 2 raised to the power of the exponent value.

Example:

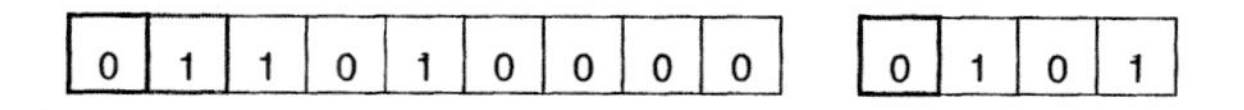

Mantissa = $2^0 + 2^{-1} + 2^{-3} = 1 + 0.5 + 0.125 = 1.625$

Exponent = $2^2 + 2^0 = 4 + 1 = 5$

Decimal Value = $1.625 \times 2^5 = 52.0$

FIGURE 3-4. Simplified binary floating-point representation, comprised of a *mantissa* (fraction part) and an *exponent.* Floating-point representations used in DSP processors, while somewhat longer and more elaborate than this example, are similar in structure. Note that the implied mantissa bit is always assumed to be 1, and therefore is never actually explicitly stored.

the mantissa is restricted to be at least 1.0. This guarantees that the precision of any floating-point value is no less than half of the maximum available precision. Thus, floating-point processors maintain very good precision with no extra effort on the part of the programmer.

In practice, floating-point DSPs generally use a 32-bit format with a 24-bit mantissa and one implied integer bit, providing 25 bits of precision. Most fixed-point DSPs use a 16-bit format, providing 16 bits of precision. So, while in theory the choice between fixed- and floating-point arithmetic could be independent of the choice of precision, in practice floating-point processors usually provide higher precision.

As mentioned above, dynamic range is defined as the ratio between the largest and smallest number representable in a given data format. It is in this regard that floating-point formats provide their key advantage. For example, consider a 32-bit fixed-point fractional representation. The minimum value that can be represented by this format is 2^{-31}; the maximum value that can be represented is $1.0 - 2^{-31}$. The ratio between these, which is the dynamic range of this data format, is approximately 2.15×10^9, or about 187 decibels (dB). A 32-bit floating-point format of the same overall size (with a 24-bit mantissa and an 8-bit exponent) can represent numbers from approximately 5.88×10^{-39} to 3.40×10^{38}, yielding a dynamic range of approximately 5.79×10^{76}, or over 1535 dB. So, while using the same number of bits as the fixed-point format, the floating-point format provides dramatically higher dynamic range.

In applications, dynamic range translates into a range of signal magnitudes that can be processed while maintaining sufficient fidelity. Different applications have different dynamic range needs. For telecommunications applications, dynamic range in the neighborhood of 50 dB is usually sufficient. For high-fidelity audio applications, 90 dB is a common benchmark. It's often helpful, though, if the processor's numeric representation and arithmetic hardware have somewhat more dynamic range than the application demands, as this frees the programmer from some of the painstaking scaling that may otherwise be needed to preserve adequate dynamic range. Scaling is discussed in more detail in Chapter 4, "Data Path."

Floating-point DSP processors are generally costlier than their fixed-point cousins, but easier to program. The increased cost results from the more complex circuitry required within the floating-point processor, which implies a larger chip. In addition, the larger word sizes of floating-point processors often means that off-chip buses and memories are wider, raising overall system costs.

The ease-of-use advantage of floating-point processors is due to the fact that in many cases the programmer does not have to be concerned about dynamic range and precision. On a fixed-point processor, in contrast, programmers often must carefully scale signals at various stages of their programs to ensure adequate numeric performance with the limited dynamic range and precision of the fixed-point processor.

Most high-volume, embedded applications use fixed-point processors because the priority is low cost. Programmers and algorithm designers determine the dynamic range and precision needs of their application, either analytically or through simulation, and then add scaling operations into the code if necessary. For applications that are less cost-sensitive, or that have extremely demanding dynamic range and precision requirements, or where ease of programming is paramount, floating-point processors have the advantage.

3.2 Native Data Word Width

The native data word width of a processor is the width of data that the processor's buses and data path can manipulate in a single instruction cycle. All common floating-point DSPs use a 32-bit native data word width. For fixed-point DSPs, the most common native data word size is 16 bits. Motorola's DSP5600x and DSP563xx families use a 24-bit data word, while Zoran's ZR3800x uses a 20-bit data word. The size of the data word has a major impact on processor cost because it strongly influences the size of the chip and the number of package pins required as well as the size and number of external memory devices connected to the DSP. Therefore, designers try to use the chip with the smallest word size that their application can tolerate. Clarkspur Design's CD2450 DSP core is unique in allowing the chip designer to select any data word width between 16 and 24 bits.

As with the choice between fixed-point and floating-point chips, there is often a trade-off between word size and development complexity. An application that appears to require 24-bit data for adequate performance can sometimes be coaxed into a 16-bit processor at the cost of more complex algorithms and/or programming.

3.3 Extended Precision

Extended precision means the use of data representations that provide higher precision than that of a processor's native data format. Extended precision can be obtained in two ways.

First, many fixed- and floating-point processors provide built-in support for an *extended precision* format for operations taking place exclusively within the data path of the processor (see Chapter 4 for a discussion of data paths). This means that as long as a series of arithmetic operations is carried out exclusively within the processor's data path and does not involve transferring intermediate results to and from memory, a data word width larger than the native data word width is available. This allows a series of arithmetic operations to be performed using extra precision and/or dynamic range, with a final rounding operation performed when the result is stored to memory.

Second, it's generally possible, though often painful, to perform *multiprecision* arithmetic by constructing larger data words out of sequences of native-width data words. For example, with a 16-bit fixed-point processor, a programmer can form 32-bit data words by stringing together pairs of 16-bit words. The programmer can implement multiprecision arithmetic operations by using the appropriate sequences of single-precision instructions. For example, a multiprecision addition can often be implemented by using a series of single-precision add instructions.

Of course, because each multiprecision arithmetic operation requires a sequence of single-precision instructions, multiprecision arithmetic is much slower than single-precision. However, some processors provide features that ease multiprecision arithmetic. These include the ability to preserve the carry bit resulting from a single-precision addition operation for use as an input into a subsequent addition, and the ability to treat multiplication operands as signed or unsigned under program control.

If the bulk of an application can be handled with single-precision arithmetic, but higher precision is needed for a small section of the code, then the selective use of multiprecision arithmetic may make sense. If most of the application requires higher precision, then a processor with a larger native data word size may be a better choice, if one is available.

3.4 Floating-Point Emulation and Block Floating-Point

Even when using a fixed-point processor, it is possible to obtain the precision and dynamic range of general-purpose floating-point arithmetic by using software routines that emulate the behavior of a floating-point processor. Some processor manufacturers provide a library of floating-point emulation routines for their fixed-point processors. If a library is not available, then the emulation routines must be written by the user. Floating-point routines are usually very expensive to execute in terms of processor cycles. This implies that floating-point emulation may be appropriate if only a very small part of the arithmetic computations in a given application require floating-point. If a significant amount of floating-point arithmetic is needed, then a floating-point processor is usually the appropriate choice.

Another approach to obtaining increased precision and dynamic range for selected data in a fixed-point processor implementation is a *block floating-point representation*. With block floating-point, a group of numbers with different mantissas but a single, common exponent is treated as a block of data. Rather than store the exponent within part of each data word as is done with general-purpose floating-point, the shared exponent is stored in its own separate data word. For example, a block of eight data values might share a common exponent, which would be stored in a separate data word. In this case, storage of an entire block of eight data values would require nine memory locations (eight for the mantissas and one for the exponent).

Block floating-point is used to maintain greater dynamic range and precision than can be achieved with the processor's native fixed-point arithmetic formats. For example, a filter routine may result in a series of 32-bit data values which the programmer then reduces to block floating-point representation using 16-bit mantissas and one four-bit exponent per block of data. Recall from our earlier discussion of floating-point formats that the value represented by a floating-point number is computed via an expression of the form:

$$value = mantissa \times 2^{exponent}$$

With block floating-point formats, the exponent for a block of data is typically determined by the data element in the block with the largest magnitude. If a given block of data consists entirely of small values, then these small data values can be shifted left several places, resulting in a relatively large negative exponent. The left shifts move lower-order bits into positions where they will be preserved when the bit width of the data is reduced. If a data block contains large data values, the data values are shifted left only a few places or none at all, and a small negative exponent (or a zero exponent) results. If the data cannot be shifted at all, then the resulting precision is the same as if the native fixed-point format were used directly. Figure 3-5 illustrates the concepts of block floating-point representations.

The conversion between the processor's native fixed-point format and block floating-point format is performed explicitly by the programmer through software. Some processors have hardware features to assist in the use of block floating-point formats. The most common of these is an "exponent detect" instruction. This instruction computes the shift needed to convert a high-preci-

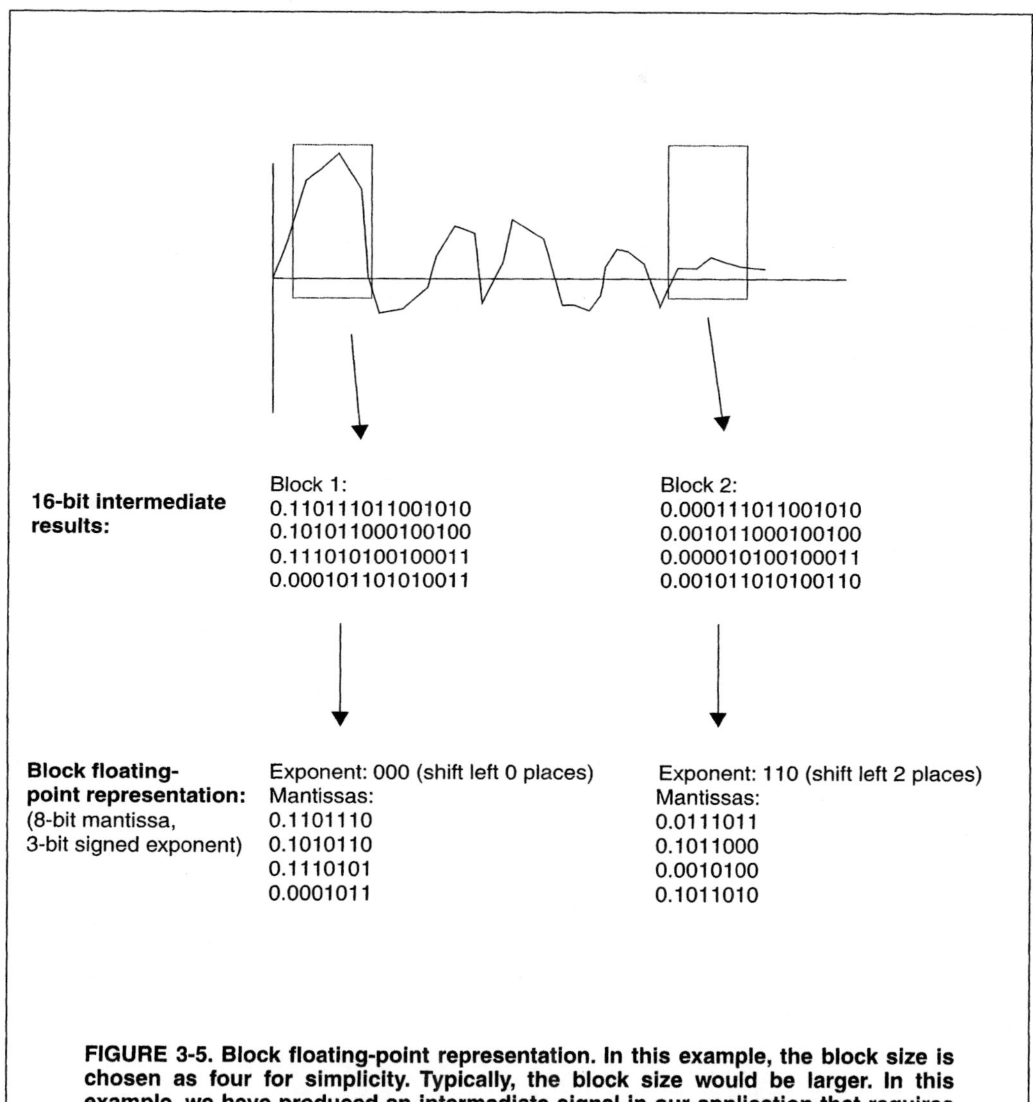

FIGURE 3-5. Block floating-point representation. In this example, the block size is chosen as four for simplicity. Typically, the block size would be larger. In this example, we have produced an intermediate signal in our application that requires 16 bits for its full representation, but we have only 8 bits available to store the samples. We use block floating-point to maintain the best precision for each block of four samples. For each block of four samples, we determine the exponent by finding the value in the block with the largest magnitude. The exponent for the block is equal to the negation of the number of left-shifts (or doublings) we can apply to this largest-magnitude value without causing overflow.

sion intermediate result (for example, a value in an accumulator) to block floating-point format. Specialized instructions that support block floating-point are discussed further in Chapter 7.

3.5 IEEE-754 Floating-Point

In 1985, the Institute of Electrical and Electronics Engineers released IEEE Standard 754 [IEE85], which defines standard formats for floating-point data representations and a set of standard rules for floating-point arithmetic. The rules specify, for example, the rounding algorithms that should be provided in a floating-point processor and how the processor should handle arithmetic exception conditions, such as divide by zero or overflow.

A few commercial DSP processors provide partial hardware support for IEEE-754 floating-point formats and arithmetic. The Motorola DSP96002 features hardware support for single-precision floating-point arithmetic as specified in IEEE-754. The Analog Devices ADSP-210xx family processors provide nearly complete hardware support for single-precision floating-point arithmetic as specified in the standard.

Some other floating-point processors, such as the AT&T DSP32xx, do not internally conform to IEEE-754, but do provide special hardware for fast conversion of numbers between the processor's internal floating-point representation and IEEE-754 representation. Hardware support for format conversion can be important in applications that require a non-IEEE-754-compliant DSP to interface with other processors that use the IEEE-754 representation. Without hardware conversion support, the noncompliant floating-point DSP must use software routines to convert between the different floating-point formats, and this software conversion can be quite time consuming. Therefore, developers of applications that require a DSP to interface with other processors that use the IEEE-754 representation should evaluate the practicality of software conversion carefully, or choose a processor with hardware conversion capabilities (or one that uses IEEE floating-point formats internally).

3.6 Relationship between Data Word Size and Instruction Word Size

While most DSP processors use an instruction word size equal to their data word size, not all do. For example, the Analog Devices ADSP-21xx family and the IBM MDSP2780 use a 16-bit data word and a 24-bit instruction word. Similarly, Zoran's 20-bit ZR3800x uses a 32-bit instruction word. Processors with dissimilar word sizes generally have provisions to allow data to be stored in program memory, for example, using the low-order 16 bits of a 24-bit program memory location. While this arrangement works, it obviously is not the most efficient use of memory since a significant portion of each program memory word used to store data is wasted, and this can impact overall system cost.

Chapter 4

Data Path

The *data path* of a DSP processor is where the vital arithmetic manipulations of signals take place. DSP processor data paths, not surprisingly, are highly specialized to achieve high performance on the types of computation most common in DSP applications, such as multiply-accumulate operations. The capabilities of the data path, along with the memory architecture (discussed in Chapter 5), are the features that most clearly differentiate DSP processors from other kinds of processors. Data paths for floating-point DSP processors are significantly different than those for fixed-point DSPs because of the differing requirements of the two kinds of arithmetic. We will first discuss fixed-point data paths and then floating-point data paths.

4.1 Fixed-Point Data Paths

Fixed-point DSP processor data paths typically incorporate a multiplier, an ALU (arithmetic logic unit), one or more shifters, operand registers, accumulators, and other specialized units. Some vendors refer to the entire data path as an ALU or arithmetic unit. In contrast, we use the term *ALU* to refer to the combination adder/subtractor/logical function unit. We use *data path* to refer to the complete arithmetic processing path, including multipliers, accumulators, and the like. Figure 4-1 illustrates a typical fixed-point data path, in this case from the Motorola DSP5600x, a 24-bit, fixed-point processor.

The data path in DSP processors is generally not used for most memory address calculations. Instead, fixed-point DSP processors usually provide a separate hardware unit for address calculation. Often called an *address generation unit*, this hardware typically performs a rich variety of address calculations, such as modulo addressing and bit-reversed addressing. The AT&T DSP32C and DSP32xx are exceptions, providing separate floating-point and fixed-point data paths. The fixed-point data path is used for address calculations as well as fixed-point arithmetic.

The wide range of specialized addressing modes found in DSP processors is one of the factors distinguishing DSPs from other kinds of processors, such as general-purpose RISC (reduced instruction set computer) processors. DSP processor address generation units and addressing modes are discussed in detail in Chapter 6.

Multiplier

The presence of a single-cycle multiplier is central to the definition of a programmable digital signal processor. Multiplication is an essential operation in virtually all DSP applications; in many applications, half or more of the instructions executed by the processor involve multiplication. Thus, virtually all DSP processors contain a multiplier that can multiply two native-sized

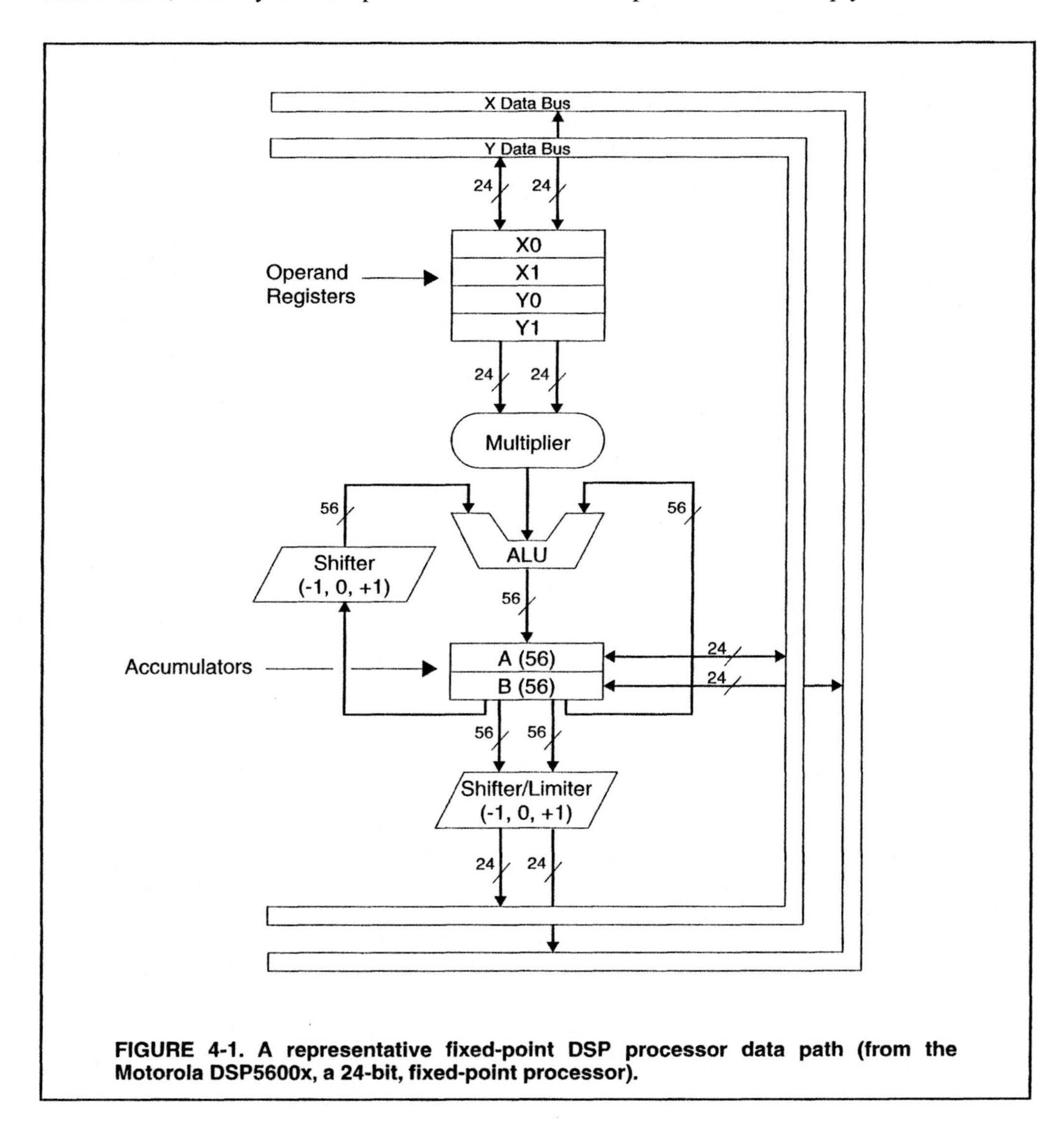

FIGURE 4-1. A representative fixed-point DSP processor data path (from the Motorola DSP5600x, a 24-bit, fixed-point processor).

operands in a single instruction cycle. Despite this commonality, there are important differences among DSP processors in terms of multiplier capabilities.

While all DSP processors are equipped with a multiplier that can produce one new result per instruction cycle, the internal *pipelining* of the multiplier can result in a delay of more than one cycle from the time inputs are presented to the multiplier until the time the result is available. (Pipelining is discussed in detail in Chapter 9.) This delay from input availability to result is called *latency*. While pipelined multipliers can produce one result every clock cycle, they achieve this performance only when long series of multiply operations are used. If a single multiply operation is preceded and followed by other kinds of operations, one or more instruction cycles must be spent waiting for the multiplier result. DSP processors with pipelined multipliers include the Clarkspur Design CD2450 DSP core.

In some cases (e.g., the Motorola DSP5600x) the multiplier is integrated with an adder to form a multiplier-accumulator unit. In other cases (e.g., the AT&T DSP16xx) the multiplier is separate; its output is deposited into a *product register*, and from there can be sent to an adder for accumulation. This distinction manifests itself in the latency suffered by a multiply-accumulate operation. If the multiplier and adder are separate, then the result of a multiply-accumulate operation is typically delayed by one instruction cycle before it can be used by the next instruction.

Another distinction among fixed-point DSP processor multipliers is the size of the product relative to the size of the operands. In general, when multiplying two n-bit fixed-point numbers, $2 \times n$ bits are required to represent the resulting product without introducing any error. This is sometimes referred to as the *law of conservation of bits*.

To understand why this is so, recall the 8-bit integer representation introduced in Figure 3-2 of Chapter 3. This format is capable of representing numbers between –128 and +127. If we multiply two large-magnitude 8-bit numbers, the result becomes too large to represent with the 8-bit format. For example, $-128 \times 128 = 16{,}384$. To represent this result, we need a format providing 16 bits.

For this reason, most fixed-point DSP processor multipliers produce a result that is twice the width of the input operands. This means that the multiplier itself does not introduce any errors into the computation. However, some fixed-point processors, for the sake of speed and economy, use multipliers that produce narrower results and thus introduce errors into some computations. For example, in the Zilog Z893xx (and the Clarkspur CD2400 core, on which the Z893xx is based), the multiplier accepts 16-bit operands, but produces a 24-bit result rather than the 32-bit result required for full precision.

Although it is possible to pass the full-width multiplication result to the next step of computation, this is usually impractical, since subsequent multiplication or additional operations would produce ever-wider results. Fortunately, in most cases it is not necessary to retain the full-width multiplication result, because there is often a point of diminishing returns beyond which additional dynamic range and precision are not useful. Therefore, for practicality's sake, the programmer usually selects a subset of the multiplier output bits to be passed on to the next computation. Or, if a series of multiplier results are to be accumulated, the accumulation may be done with the full-width results, and then the width of the final result reduced before proceeding to the next stage of computation.

Assuming that the full-width multiplication or accumulation result is not going to be retained, the programmer must select which portion of the result to retain. This selection is usually made so that the radix point of the selected result portion lies in the same position as the radix point in the normal fixed-point format being used throughout the program. This allows the selected multiplier output to be treated as any other operand in the next step of processing.

When two binary numbers are multiplied, the position of the radix point in the full-width result depends on the position of the radix points in the operands. This is illustrated in Figure 4-2 and Figure 4-3. If integer arithmetic is used, then the full-width multiplier output is also an integer, and the programmer typically retains the least-significant bits of the result. In this case, the size of operands must be constrained by the programmer so that the multiplication result fits completely in the lower half of the multiplier result. When this is done, the most significant $N/2+1$ bits of the N-bit result are all equal to the value of the sign bit (they are sometimes called *sign exten-*

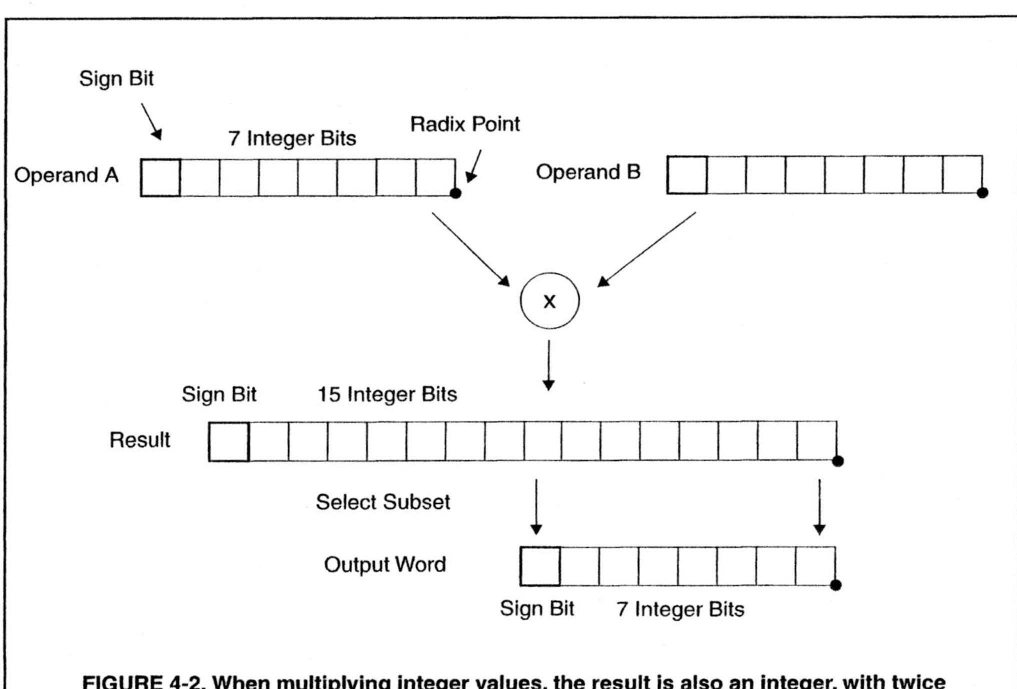

FIGURE 4-2. When multiplying integer values, the result is also an integer, with twice the overall width of the input operands. Normally, the programmer selects the least significant half of the result word for use in the next step of processing. In this case, the programmer must ensure that the full result value is contained within the lower half of the result word to prevent overflow. This can be done by constraining the size of the operands or by scaling the result. When the operands are constrained or the results have been scaled properly, the most significant $N/2+1$ bits of the N-bit result are all equal to the value of the sign bit, so no information is lost when the upper $N/2$ bits are discarded (leaving one sign bit in the final output word).

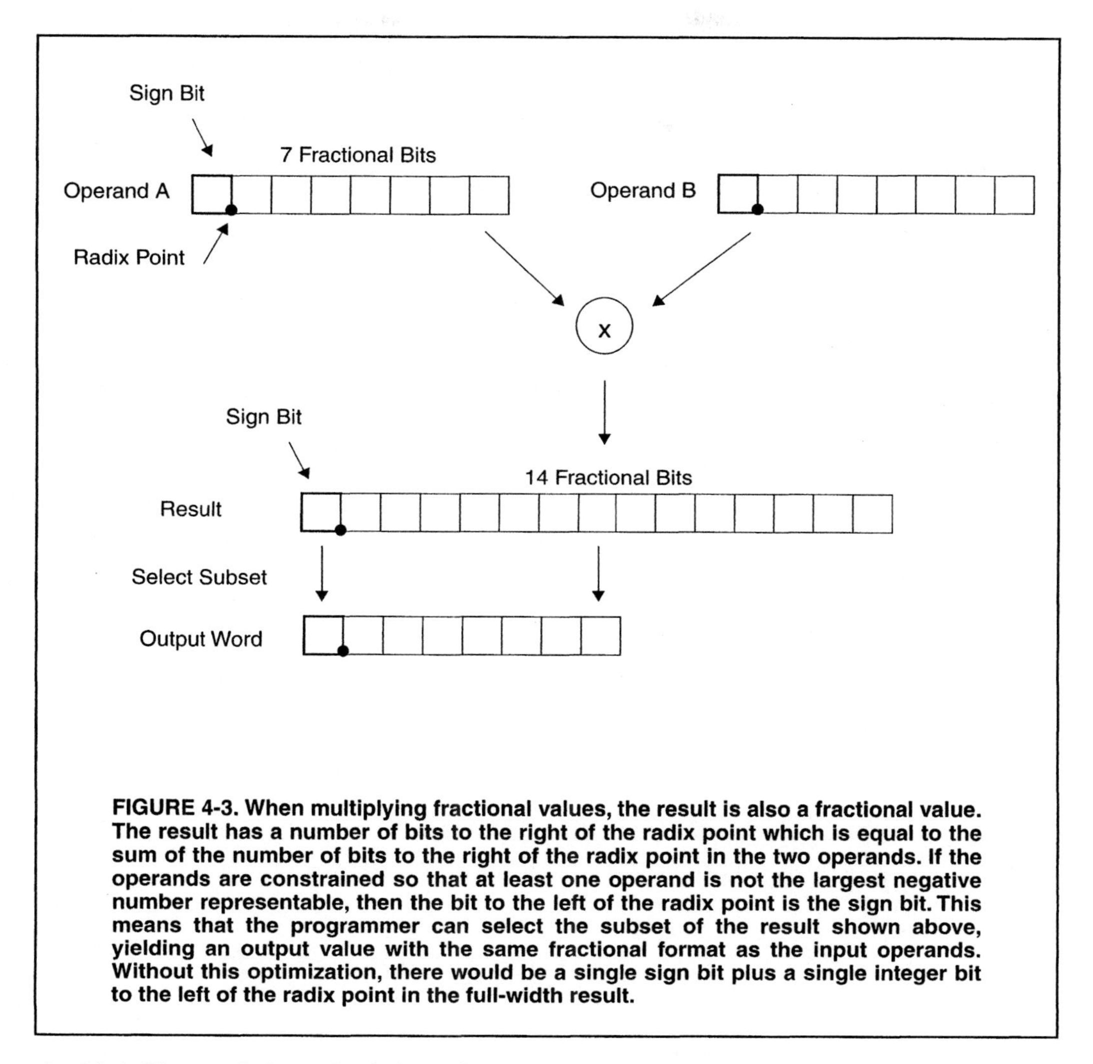

FIGURE 4-3. When multiplying fractional values, the result is also a fractional value. The result has a number of bits to the right of the radix point which is equal to the sum of the number of bits to the right of the radix point in the two operands. If the operands are constrained so that at least one operand is not the largest negative number representable, then the bit to the left of the radix point is the sign bit. This means that the programmer can select the subset of the result shown above, yielding an output value with the same fractional format as the input operands. Without this optimization, there would be a single sign bit plus a single integer bit to the left of the radix point in the full-width result.

sion bits). Thus, no information is lost when the upper $N/2$ bits are discarded (leaving one sign bit in the final output word).

If fractional arithmetic is used, then the full-width multiplier result has twice as many bits to the right of the radix point as the multiplier operands, and the programmer typically discards the least significant half of these bits, perhaps after rounding.

The difference in which part of the multiplier result is selected for use in the next step of computation is the primary difference between how integer arithmetic and fractional arithmetic are handled in fixed-point DSPs.

A common optimization employed with fixed-point multiplication is to prohibit one of the operands from taking on the largest negative value that can be represented by the numeric format being used (this value is sometimes called NMAX). If this is done, the size of the full-width result is reduced by one bit. That is, when two n-bit binary numbers are multiplied, if at least one of them is not the largest negative number representable, then the bit width of the result is $2n - 1$ instead of $2n$ bits. As shown in Figure 4-3, when fractional arithmetic is used, this optimization results in an output word that contains a single sign bit and no integer bits to the left of the radix point. This allows the programmer to select a subset of the full-width result that yields an output word with the same format as the input operands. Without this optimization, there would be a single sign bit plus a single integer bit to the left of the radix point in the full-width result. The programmer would then have to scale the full-width result by one-half before being able to select an output word with the same format as the input operands. Although the processor may allow this scaling to be done without extra instructions, the scaling would have the effect of reducing the dynamic range of the scaled data.

Fixed-point DSP processors often provide specific hardware features to help the programmer efficiently select the subset of the multiplier output bits desired. One common feature is to treat the $2n$-bit-wide multiplier output register or accumulator register as two independently addressable n-bit-wide registers. This allows the programmer to directly select the most-significant (for fractional multiplication) or least-significant (for integer multiplication) halves of the multiplier or accumulator output values to pass to the next step of computation. Another common feature is the inclusion of an automatic left shift by one bit following the multiplier. In cases where the bit-width optimization discussed above is used for fractional multiplication, this left-shift operation aligns the multiplier output so that the desired subset of the full-width result fits completely within the upper half of the result register. If the upper half of the result register is independently addressable, then the programmer can select the desired bits of the full-width result simply by reading out the upper half of the result register. Without this built-in left shift, the programmer would have to perform an explicit shift operation before reading out the desired portion of the result. In some processors, this left shift can be enabled or disabled through the use of a mode bit, simplifying switching between fractional and integer arithmetic.

Accumulator Registers

Accumulator registers hold intermediate and final results of multiply-accumulate and other arithmetic operations. Most DSP processors provide two or more accumulators. A few processors provide only a single accumulator, which can be a drawback for many applications. When only one accumulator is available, the accumulator often becomes a bottleneck in the architecture: since the accumulator is usually used as one of the source operands and as the destination operand for ALU operations, its contents must be loaded or stored frequently as the ALU is used for various tasks. These loads and stores limit the rate at which data can flow through the ALU.

Ideally, the size of the accumulator registers should be larger than the size of multiplier output word by several bits. The extra bits, called *guard bits*, allow the programmer to accumulate a number of values without the risk of overflowing the accumulator and without the need for scaling intermediate results to avoid overflow. An accumulator with n guard bits provides the capacity for up to 2^n values to be accumulated without the possibility of overflow. Most processors provide either four or eight guard bits. For example, the AT&T DSP16xx provides four guard bits (36-bit accumulators with a 32-bit multiplier product), while the Analog Devices ADSP-21xx provides eight guard bits (40-bit accumulators with a 32-bit multiplier product).

On a processor that lacks guard bits, input signals or intermediate results often must be scaled before being added to the accumulator if the possibility of overflow is to be eliminated. Usually this involves scaling the multiplier result by shifting it right by a few bits. Some processors that do not provide guard bits in the accumulator are capable of shifting the product register value before adding it to the accumulator without requiring additional instruction cycles. For example, the Texas Instruments TMS320C2x and TMS320C5x allow the product register to be automatically shifted right by six bits. As described in the discussion of shifters below, such scaling results in a loss of precision. However, unless the amount of scaling used is extreme or the number of products being accumulated is very large, the loss of precision introduced by scaling the product is small.

A further argument in favor of shifting the product right before accumulation is that when fractional arithmetic is used, often only the most significant half of the accumulator is retained after a series of multiply-accumulates, as discussed earlier. In this case, the loss of precision of intermediate values due to scaling usually does not affect the final result. This is because the quantization error present in the result due to scaling is usually entirely contained in the discarded least-significant half of the accumulator.

Guard bits provide greater flexibility than scaling the multiplier product because they allow the maximum precision to be retained in intermediate steps of computation. However, support for scaling the multiplier result in lieu of guard bits is sufficient for many applications. A few processors, such as the Texas Instruments TMS320C1x, lack both guard bits and the ability to efficiently scale the product register. This requires the multiplier input to be scaled to avoid overflow, which can result in significantly reduced precision. The lack of both accumulator guard bits and support for scaling the product register is a serious limitation in many situations.

ALU

DSP processor *arithmetic logic units* implement basic arithmetic and logical operations in a single instruction cycle. Common operations include add, subtract, increment, negate, and logical *and*, *or*, and *not*. ALUs differ in the word size they use for logical operations. Some processors perform logical operations on operands that are the full width of the accumulator, while others can perform logical operations only on native-width data words. For example, the AT&T DSP16xx performs logical operations on 36-bit accumulator values, while the Motorola DSP5600x, which has a 56-bit accumulator, performs logical operations on 24-bit data words. If the ALU cannot perform logical operations on accumulator-width data, then programmers needing this capability must resort to performing logical operations in multiple steps, which complicates programming and consumes instruction cycles.

As mentioned above, in some processors the ALU is used to perform addition for multiply-accumulate operations. In other processors, a separate adder is provided for this purpose.

Shifter

Multiplication and accumulation tend to result in growth in the bit width of arithmetic results. In most cases, the programmer will want to choose a particular subset of the result bits to pass along to the next stage of processing. A shifter in the data path eases this selection by scaling (multiplying) its input by a power of two (2^n).

Scaling is an important operation in many fixed-point DSP applications. This is because many basic DSP functions have the effect of expanding or contracting the range of values of the signals they process. Consider the simple example of a filter, as illustrated in Figure 4-4. The example filter has a *gain* of 100. This means that the range of values at the output of the filter can be as much as 100 times larger than the range of values at the input to the filter. If the input signal is limited to the range –1.0 to +1.0, the output values are limited to the range –100 to +100. The difficulty arises because the numeric representation has a limited range, normally –1.0 to +1.0 for fractional representations. If signals exceed these values, overflow occurs, and incorrect results are produced. To avoid this situation, the programmer must be aware of the range of signal values at each point in the program and scale signals at various points to either eliminate the possibility of overflow or reduce the probability of overflow to an acceptably low level. As shown in Figure 4-4(b), the programmer could eliminate the possibility of overflow by scaling the signal x_n by a factor of 0.0078 (1/128, the nearest power of 2 to 1/100) before filtering it.

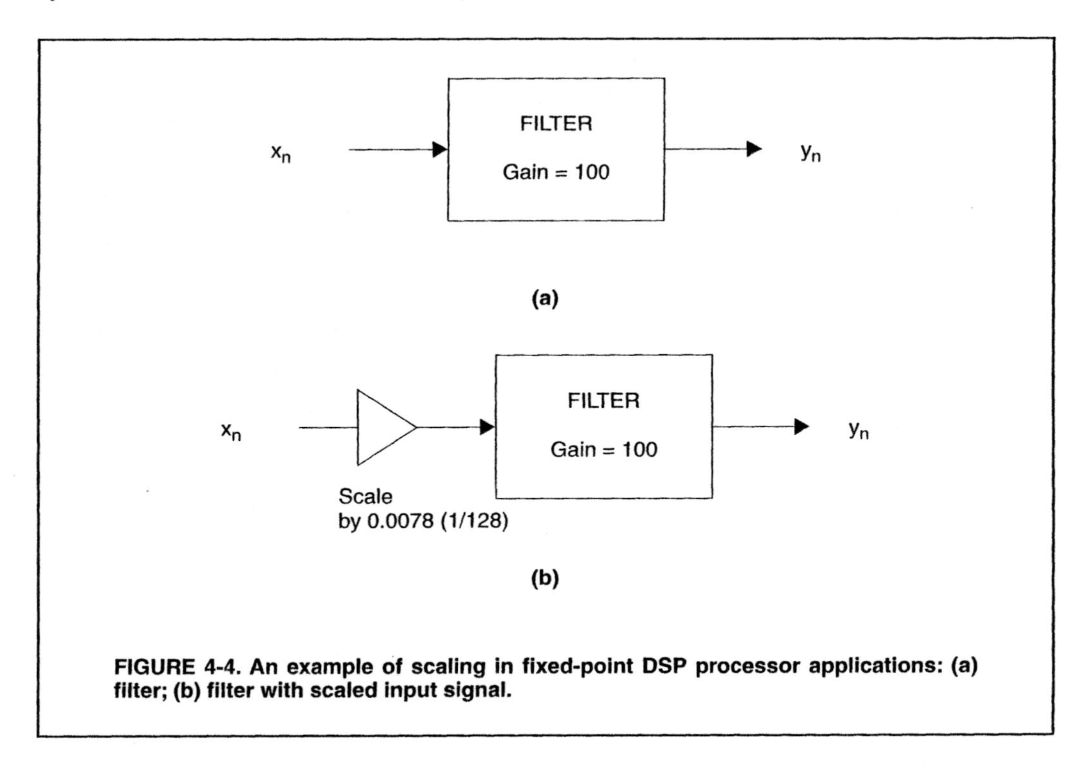

FIGURE 4-4. An example of scaling in fixed-point DSP processor applications: (a) filter; (b) filter with scaled input signal.

The trade-off that comes with scaling signals in this way is the loss of precision and dynamic range when a signal is scaled to a smaller range. Recall that precision is the ratio between the magnitude of the value being represented and the maximum possible magnitude of the quantization error of the representation. When a signal is scaled to a smaller range, some of the lower-order bits of the original value are lost through truncation or rounding. This means that the magnitudes of the values represented are reduced, while the maximum possible magnitude of the quantization error is unchanged. Therefore, precision is reduced. Similarly, since the magnitude of the largest representable value is reduced, but the magnitude of the smallest representable value is unchanged, so dynamic range is reduced.

Thus, scaling must be done with great care, balancing the need to reduce or eliminate the possibility of overflow with the need to maintain adequate dynamic range and precision. Proper scaling of signals can be a significant challenge in implementing an application on a fixed-point processor. Typically, application developers use detailed simulations and analysis to determine where and by how much signals should be scaled and then insert scaling operations at the appropriate locations.

Note that scaling is related to accumulator guard bits in that both are used to eliminate or reduce the possibility of overflow. While scaling limits the range of intermediate or final results to the range of representable values, guard bits provide an intermediate representation that has a larger dynamic range. As described above, scaling of intermediate results can sometimes be used to remove the need for guard bits. However, guard bits do not remove the need for scaling. When guard bits are in use it is necessary to scale the final result in order to convert from the intermediate representation to the final one. For example, in a 16-bit processor with four guard bits in the accumulator, it may be necessary to scale accumulator values by 2^{-4} before writing them to memory as 16-bit values.

A shifter is often found immediately following the multiplier and ALU. Some processors provide a shifter between the multiplier and ALU to allow scaling the product as discussed above. Some shifters have limited capabilities, for example, offering a left shift by one bit (scale by 2^1), a right shift by one bit (scale by 2^{-1}), or no shift. Such shifters can perform multibit shifts one bit at a time, but this can be time consuming. Another kind of shifter, called a *barrel shifter*, offers more flexibility by supporting shifts by any number of bits in a single instruction cycle. In some processors, a limited-capability shifter is located immediately after the multiplier, so that multiplication results can be scaled before being passed to the accumulator or ALU for further processing.

Some DSP processors provide multiple shifters with different capabilities in different places in the data path. This allows greater flexibility to perform shifting in the most appropriate places for each algorithm being implemented. For example, the DSP5600x has two independent, limited-capability shifters: one is used to scale multiply-accumulate results as they are written from the accumulator to memory, and the other is used to shift values within the accumulator, including logical (no sign extension) and rotate-type shifts.

Overflow and Saturation

Many DSP applications involve accumulating (adding together) a series of values. This happens, for example, in filtering algorithms, where elements of a data series are multiplied by coefficients and the resulting products are summed together. When a series of numbers is accumulated, the magnitude of the sum may grow. Eventually, the magnitude of the sum may exceed the maximum value that can be represented by the accumulator register. In this situation, called *overflow*, an incorrect value is stored. When overflow occurs, the actual output of the accumulator can be very far from the correct value.

To understand the effects of overflow, consider adding base-10 numbers in a system where numbers cannot be larger than two digits in size. If we add the numbers 50, 45, and 20, the result is 15, because two digits are not sufficient for representing the correct result of 115.

Even if the accumulator register does not overflow, overflow can still occur when the accumulated value is transferred to memory if the accumulator provides guard bits. In this case, the accumulator can represent larger numbers than can be stored in a single word in memory. This is the case in many DSP processors, including DSP Group's PineDSPCore and AT&T's DSP16xx. Overflow can also occur if a shifter is used to scale up the accumulator value as it is stored to memory.

There are two common ways of dealing with overflow. The first technique is to carefully scale all computations to eliminate the possibility of overflow, regardless of the input data. This can be effective, but it may require that signals be scaled to such small values that adequate signal fidelity cannot be maintained. An alternative is to use *saturation arithmetic*. In saturation arithmetic, a special circuit detects when overflow has occurred and replaces the erroneous output value with the largest positive number that can be represented (in the case of overflow in the positive direction) or with the largest negative number that can be represented (in the case of overflow in the negative direction). The result, of course, is still incorrect, but the error is smaller than it would be without saturation. Referring again to our 2-digit decimal calculator, the result of adding 50, 45, and 20 without saturation is 15, which is 100 away from the correct result of 115. With saturation, the result is 99 (the largest positive number that we can represent with two digits), which is only 16 away from the correct result.

Because it often is not practical or desirable to scale signals to eliminate the possibility of overflow, saturation arithmetic is very useful. Fixed-point DSP processors generally provide special hardware for saturation arithmetic, so that it occurs automatically (perhaps under the control of a mode register) or with the execution of a special instruction. The hardware unit that implements saturation arithmetic is called a *limiter* by some manufacturers.

Rounding

As discussed above, multiplication, accumulation, and other arithmetic operations tend to increase the number of bits needed to represent arithmetic results without loss of precision. At some point, it is usually necessary to reduce the precision of these results (for example, when transferring the contents of a 36-bit accumulator into a 16-bit memory location). The simplest way to do this is to discard the least significant bits of the representation. This operation is called *truncation*. For example, to truncate a 36-bit value to 16 bits, the least significant 20 bits can be discarded, leaving only the most significant 16 bits. Since the information contained in the discarded bits is lost, an error is introduced to the signal. Note that the truncated value is always smaller than or equal to the original. This means that truncation adds a *bias* or offset to signals. *Rounding* techniques reduce the arithmetic error as well as the bias introduced by this reduction in precision. As with saturation, some processors perform rounding automatically when results are transferred between certain registers and memory (this may be enabled by a mode register). Other processors provide a special instruction for rounding.

The simplest kind of rounding is the so-called *round-to-nearest* technique. This is the conventional type of rounding that we use in everyday arithmetic, and it is the type of rounding provided by most fixed-point DSP processors. With this approach, numbers are rounded to the nearest value representable in the output (reduced-precision) format; numbers that lie exactly at the midpoint between the two nearest output values are always rounded up to the higher (more positive) output value.

The standard way of implementing this scheme is to add a constant equal to one half the value of the least significant bit of the output word to the value to be rounded, and then truncate the result to the desired width. Some processors provide a special instruction to facilitate rounding by preloading the accumulator with the appropriate constant value. Without such an instruction, round-to-nearest rounding can be performed by using a normal move instruction to preload the appropriate constant into the accumulator, or by adding the constant to the accumulator prior to truncation.

The fact that round-to-nearest rounding always rounds numbers that lie exactly at the midpoint between the two nearest output values to the higher output value introduces an asymmetry. If we round a typical set of signal values in this way, on average more values are rounded up than are rounded down. This means that the round-to-nearest operation adds a bias to signals. This bias is typically many orders of magnitude smaller than the bias introduced by truncation. However, for some applications, such as IIR (infinite impulse response) filters and adaptive filters, this small bias can still be troublesome. Round-to-nearest rounding has the advantage of being simple to implement.

An improved rounding scheme that addresses the bias problem of the round-to-nearest approach is *convergent* rounding. Convergent rounding is slightly more sophisticated than the familiar round-to-nearest technique. In convergent rounding, when a number to be rounded lies exactly at the midpoint between the two nearest output values, it may be rounded higher or lower. The rounding direction depends on the value of the bit of the number that is in the position that will become the least significant bit (LSB) of the output word. If this bit is a zero, then the number is rounded down (in the negative direction); if this bit is a one, the number is rounded up. Convergent rounding is compared to the round-to-nearest technique in Figure 4-5.

For most signals, the bit that is used to decide whether to round midpoint cases up or down is assumed to be equally likely to be zero or one. Therefore, the convergent mechanism rounds up in half of the midpoint cases and down in half. If this assumption holds, then convergent rounding effectively avoids the bias caused by the round-to-nearest approach. Even though it is relatively simple to implement, convergent rounding is not supported in hardware by most fixed-point DSPs. The Analog Devices ADSP-21xx and Motorola DSP5600x families are two that do provide convergent rounding. Processors that support convergent rounding can also perform conventional rounding using the techniques described above.

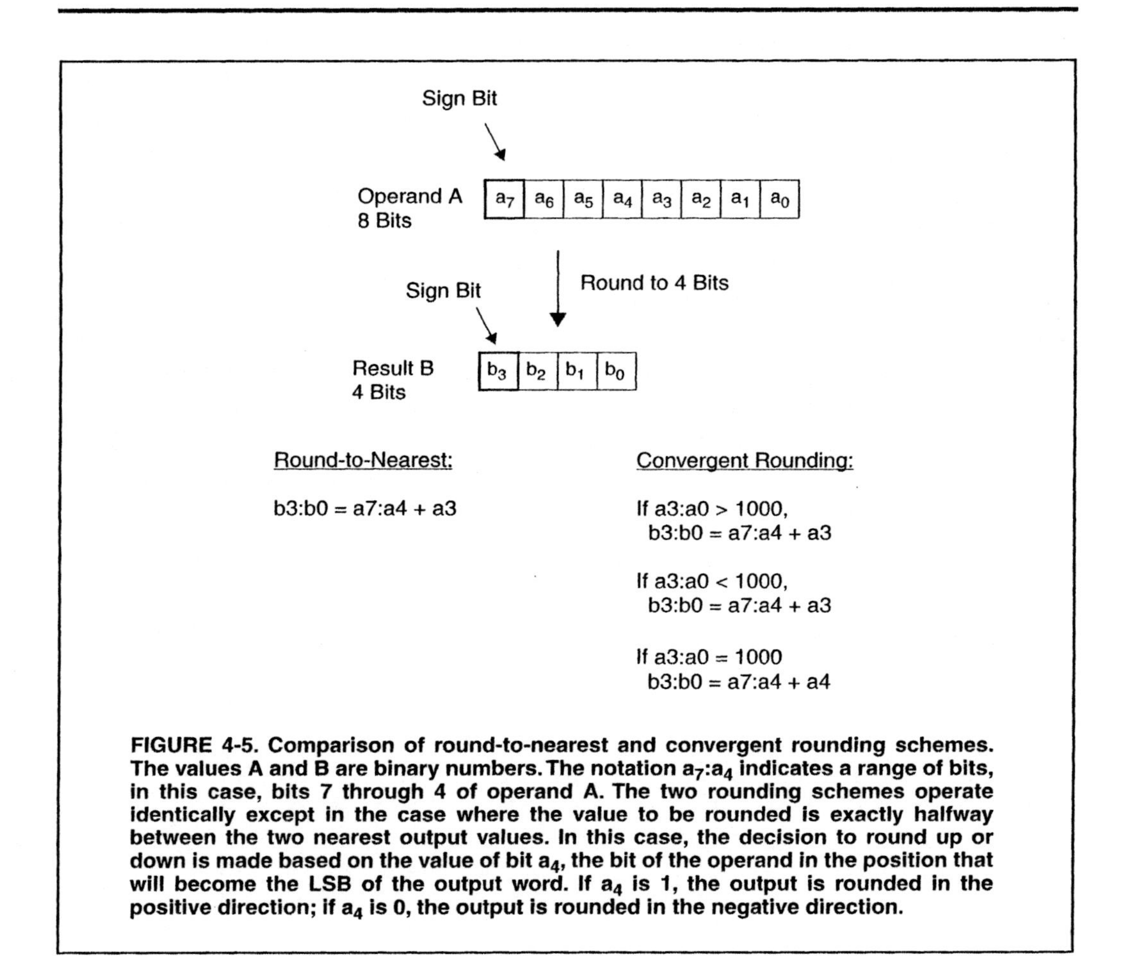

FIGURE 4-5. Comparison of round-to-nearest and convergent rounding schemes. The values A and B are binary numbers. The notation a_7:a_4 indicates a range of bits, in this case, bits 7 through 4 of operand A. The two rounding schemes operate identically except in the case where the value to be rounded is exactly halfway between the two nearest output values. In this case, the decision to round up or down is made based on the value of bit a_4, the bit of the operand in the position that will become the LSB of the output word. If a_4 is 1, the output is rounded in the positive direction; if a_4 is 0, the output is rounded in the negative direction.

The error introduced by rounding and truncation in DSP algorithms is often modeled as random noise or *quantization noise*. Note that the quantization noise due to truncation, round-to-nearest rounding, and convergent rounding all have the same noise power. The key difference between these three techniques for reducing precision is the bias they add to the signal.

In some applications (especially in telecommunications) the rounding technique to be used is specified by a published technical standard.

Operands and Operand Registers

In most fixed-point DSP processors, operands for the data path are supplied from a small number of operand registers or from the accumulator(s) within the data path. Values must be loaded into the operand registers with explicit *move* instructions before they can be processed by the data path. A processor that processes data only after loading it into operand registers is often referred to as having a *load-store architecture*. Registers are accessed using *register-direct* addressing. (Refer to Chapter 6 for a detailed discussion of addressing modes.)

On some processors (for example, the Texas Instruments TMS320C2x/C5x and the PineDSPCore from DSP Group), operands can be supplied directly from memory to the data path, using memory-direct or register-indirect addressing.

4.2 Floating-Point Data Paths

Floating-point DSP processor data paths are similar to those found in fixed-point DSP processors, but differ in several respects. In this section, we emphasize those aspects of floating-point DSP data paths that are different from their fixed-point cousins. Figure 4-6 illustrates a typical floating-point data path, in this case from the AT&T DSP3210, a 32-bit, floating-point processor.

In most floating-point DSPs, the main data path is capable of both floating-point and fixed-point calculations, but can handle only one type of operation in a given instruction cycle. This is the case with the floating-point DSPs from Texas Instruments, Analog Devices, and Motorola.

Some floating-point processors provide two data paths: one for floating-point operations and one (with lesser capabilities) for fixed-point operations. This is true of the AT&T DSP32xx, for example, where the control arithmetic unit handles both address calculations and general-purpose integer arithmetic, while the data arithmetic unit handles floating-point calculations. In processors that have separate fixed-point and floating-point data paths, the fixed-point data path does not include a multiplier.

Multiplier

Floating-point DSP multipliers accept two native-size (usually 32-bit) floating-point operands. Unlike fixed-point DSP multipliers, floating-point multipliers generally do not produce an output word large enough to avoid loss of precision. For example, when multiplying two IEEE-754 single-precision floating-point values, each input value has an effective mantissa width of 24 bits. To maintain full precision, the output value requires an effective mantissa width of 48 bits. Most floating-point DSPs do not accommodate this full width. Instead, the output format supported by floating-point DSP multipliers is commonly somewhat larger than the input format, usually providing an extra eight to twelve bits of mantissa precision.

ALU

Floating-point DSP processor ALUs typically provide addition, subtraction, and other arithmetic operations such as absolute value, negate, minimum, and maximum. Some ALUs provide other specialized operations, examples of which are shown in Table 4-1.

Floating-point processors use their ALUs to perform addition for multiply-accumulate operations. In addition to multiply-accumulate operations, some processors (for example, the AT&T DSP32C) provide a multiply-add operation. The multiply-add operation is distinguished from the multiply-accumulate in that the result is written into a different accumulator than the one that provides the addend value for addition with the product.

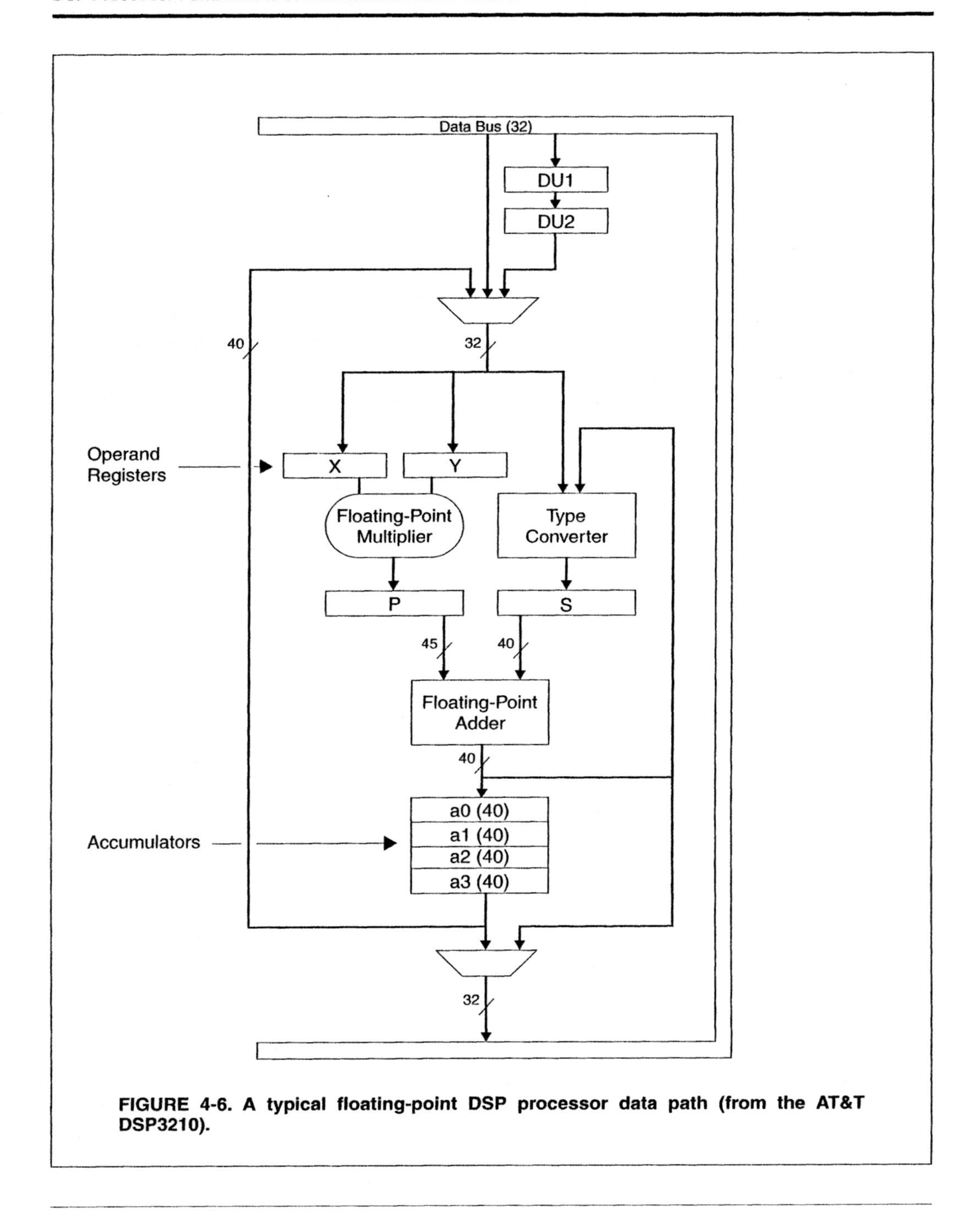

FIGURE 4-6. A typical floating-point DSP processor data path (from the AT&T DSP3210).

TABLE 4-1. Examples of Specialized ALU Operations Found in Some Floating-Point DSPs

Operation	Description	Example Processors
Reciprocal seed	Provides an estimate of the reciprocal of a value (i.e., $1/x$). The estimate can be used as the starting point, or *seed*, for an iterative algorithm that computes the precise value of the reciprocal.	ADSP-210xx DSP96002 DSP32xx
Square root reciprocal seed	Provides an estimate of the reciprocal of the square root of a value. The estimate can be used as the starting point, or *seed*, for an iterative algorithm that computes the precise value of the reciprocal of the square root.	ADSP-210xx DSP96002
Add and subtract	Simultaneously computes the sum and the difference of two data values and deposits the results into two registers.	ADSP-210xx DSP96002
IEEE format conversion	Converts between the processor's internal floating-point data format and IEEE-754 floating-point format.	DSP32xx TMS320C4x
Integer floating-point conversion	Converts between internal floating-point and internal integer formats.	DSP32xx TMS320C3x TMS320C4x

Bit-wise logical operations such as *and*, *or*, and *not* are generally not meaningful when applied to floating-point data and are usually not provided by floating-point-only ALUs.

Overflow, Underflow, and Other Exceptions

With floating-point arithmetic, overflow and other conditions like it are often called *exceptions*. Exceptions are (hopefully) unusual conditions that may cause erroneous arithmetic results. The processor indicates an exception has occurred by setting the appropriate bit in a status register or by causing an interrupt (optional on some processors) which then triggers an exception-handling routine supplied by the programmer.

Overflow is much less of a concern with floating-point DSP processors than with fixed-point DSPs because of the much larger dynamic range provided by the floating-point formats (see Chapter 3 for details). However, for some applications overflow may still be a real concern. Most floating-point DSP processors record the occurrence of overflow by setting a status flag and automatically saturating the result of the operation that caused the overflow (setting it to the largest positive or largest negative output value representable).

Another exception condition tracked by floating-point processors is *underflow*. Underflow occurs when the result of an arithmetic operation is too small to be represented. This can happen, for example, when multiplying two tiny numbers together. In this case, the usual action is for the processor to set the result to zero and to indicate that underflow has occurred by setting a flag in a status register.

Other exception conditions are also monitored by some processors. For example, division by zero causes an exception for processors that provide a reciprocal seed operation.

Rounding

As with fixed-point arithmetic, floating-point multiplication, accumulation, and other arithmetic operations tend to increase the bit width of arithmetic results. Most floating-point DSP processors deal with this growth in precision by automatically rounding arithmetic results to a 40- to 44-bit intermediate format. This intermediate format is used for computations that take place within the data path. When results are written to memory, they can be written either as extended-precision values (which requires multiple memory locations and multiple register-to-memory move operations per data value) or they can be rounded to the native single-precision format, usually 32 bits.

As with fixed-point processors, most floating-point processors provide the simplest kind of rounding, *round-to-nearest*. Some processors provide two or more options for rounding floating-point results. The Motorola DSP96002 is unique in that it provides all of the rounding modes specified by IEEE standard 754 for single-precision values: convergent rounding, round toward positive infinity, round toward zero, and round toward negative infinity.

Accumulator Registers

In general, floating-point processors have more and larger registers than their fixed-point counterparts. In some floating-point processors (for example, the AT&T DSP3210) a small number of registers are specifically designed for use as accumulators. Other processors provide a bank of general-purpose registers, some subset of which can receive the results of multiply-accumulate or other arithmetic operations. This is the case, for example, with the Texas Instruments TMS320C3x, which has eight 40-bit and eight 32-bit registers, subsets of which can receive the results of arithmetic operations.

Shifter

As with fixed-point arithmetic, a floating-point multiply-accumulate operation tends to result in growth in the bit width of arithmetic results. However, with floating-point arithmetic the hardware automatically scales the results to preserve the maximum precision possible. This is the key advantage of floating-point arithmetic, as described in detail in Chapter 3. Floating-point data paths incorporate a shifter to perform this scaling, but the shifter is generally not visible to or explicitly controllable by the programmer for floating-point operations. In processors where a single data path performs both fixed- and floating-point arithmetic, the shifter can be explicitly controlled by the programmer for shifting fixed-point data.

Operands and Operand Registers

As with fixed-point DSPs, most floating-point DSP processors supply operands for the data path from operand registers or from the accumulators within the data path. This is called *register-direct* addressing. (Refer to Chapter 6 for a detailed discussion of addressing modes.) Values must be loaded from memory into the operand registers with explicit *move* instructions. (Some processors can perform these move instructions in parallel with arithmetic instructions—see Chapter 7 for more information on parallel instructions.)

A few processors (for example, the AT&T DSP32xx) supply operands directly from memory to the data path using register-indirect addressing. Arithmetic results are written to an accumulator and may optionally also be written to memory using register-indirect addressing.

4.3 Special Function Units

As demand grows for DSP processor use in specialized applications, processor manufacturers have begun to incorporate specialized hardware into their processors' data paths to improve performance. An early example of this is the AT&T DSP1610. This processor is intended for digital mobile radio applications, where DSP processors are used in speech coding. Speech coding applications typically involve a large number of bit field operations (for example, inserting a series of n bits at a specified location within an m-bit word). The base architecture from which the DSP1610 is derived (the DSP1600) does not provide a barrel shifter and thus cannot efficiently perform bit manipulation operations. AT&T addressed this deficiency by adding a specialized bit manipulation unit (BMU) into the data path of the DSP1610. The addition of the BMU gives the DSP1610 comparable performance on bit manipulation operations to other processors that provide a barrel shifter in the main data path. The key innovation introduced by AT&T is a base architecture which accommodates the addition of application-specific function units along with instructions to access these function units.

Chapter 5

Memory Architecture

As we explored in the previous chapter, DSP processor data paths are optimized to provide extremely high performance on certain kinds of arithmetic-intensive algorithms. However, a powerful data path is, at best, only part of a high-performance processor. To keep the data path fed with data and to store the results of data path operations, DSP processors require the ability to move large amounts of data to and from memory quickly. Thus, the organization of memory and its interconnection with the processor's data path are critical factors in determining processor performance. We call these characteristics the *memory architecture* of a processor, and the kinds of memory architectures found in DSP processors are the subject of this chapter. Chapter 6 covers addressing modes, which are the means by which the programmer specifies accesses to memory.

To understand the need for large memory bandwidth in DSP applications, consider the example of a finite impulse response (FIR) filter, shown in Figure 5-1. Although this example has become somewhat overused in DSP processor circles, it is perhaps the simplest example that clearly illustrates the need for several special features of DSP processors.

The mechanics of the basic FIR filter algorithm are straightforward. The blocks labeled D in Figure 5-1 are *unit delay* operators; their output is a copy of the input sample delayed by one sample period. A series of storage elements (usually memory locations) are used to simulate a series of these delay elements (called a *delay line*). The FIR filter is constructed from a series of *taps*. Each tap includes a multiplication and an accumulation operation. At any given time, $n - 1$ of the most recent input samples reside in the delay line, where n is the number of taps in the filter. Input samples are designated x_k; the first input sample is x_1, the next is x_2, and so on. Each time a new input sample arrives, the previously stored samples are shifted one place to the right along the delay line, and a new output sample is computed by multiplying the newly arrived sample and each of the previously stored input samples by the corresponding *coefficient*. In the figure, coefficients are represented as c_n, where n is the coefficient number. The results of each multiplication are summed together to form the new output sample, y_k.

As we discussed in Chapter 4, DSP processor data paths are designed to perform a multiply-accumulate operation in one instruction cycle. This means that the arithmetic operations required for one tap can be computed in one instruction cycle. Therefore, a new output sample can be produced every n instruction cycles for an n-tap FIR filter. However, to achieve this performance, the processor must be able to make several accesses to memory within one instruction cycle. Specifically, the processor must:

- Fetch the multiply-accumulate instruction
- Read the appropriate data value from the delay line
- Read the appropriate coefficient value
- Write the data value to the next location in the delay line to shift data through the delay line

Thus, the processor must make four accesses to memory in one instruction cycle if the multiply-accumulate operation is to execute in a single instruction cycle. In practice, some processors use other techniques (discussed later) to reduce the actual number of memory accesses needed to three or even two. Nevertheless, all processors require multiple memory accesses within one instruction cycle to compute an FIR filter at a rate of one tap per instruction cycle. This level of memory bandwidth is also needed for other important DSP algorithms besides the FIR filter. Note that for clarity of explanation, in this section we ignore issues of pipelining. Pipelining is explored in detail in Chapter 9.

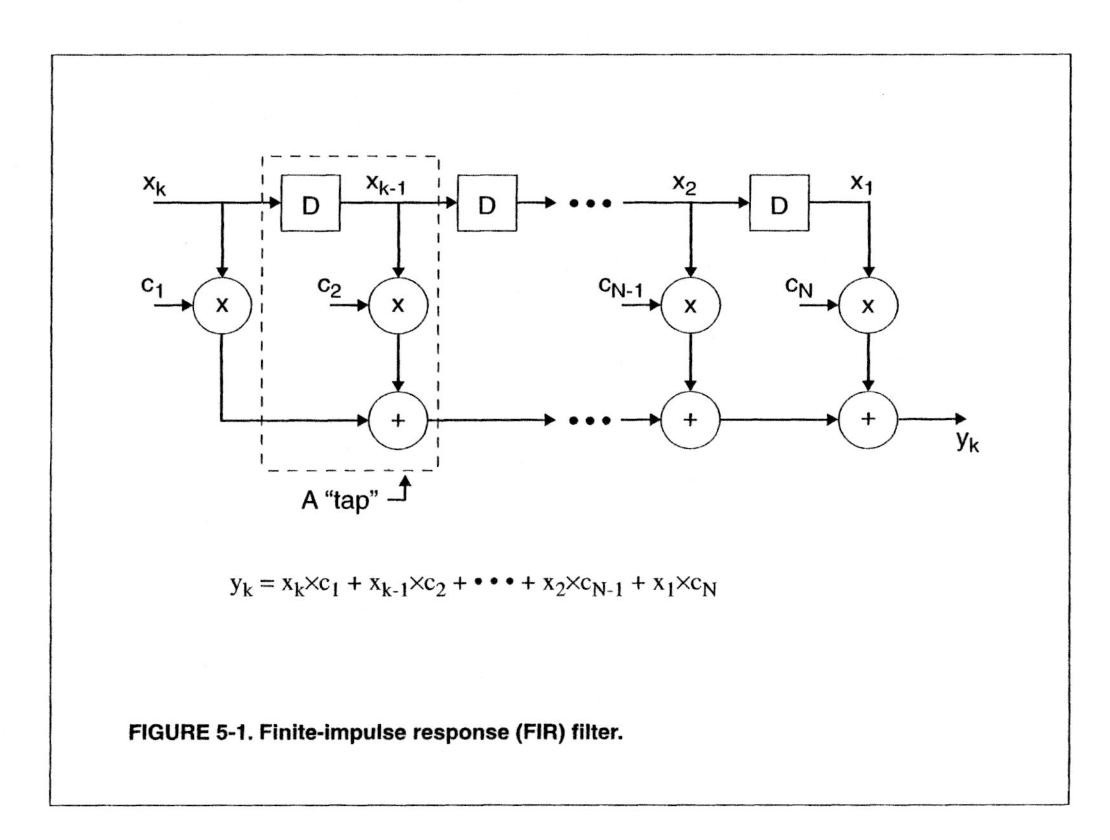

FIGURE 5-1. Finite-impulse response (FIR) filter.

5.1 Memory Structures

The simplest processor memory structure is a single bank of memory, which the processor accesses through a single set of address and data lines, as shown in Figure 5-2. This structure, which is common among non-DSP processors, is often called a *Von Neumann architecture*. Both program instructions and data are stored in the single memory. In the simplest (and most common) case, the processor can make one access (either a read or a write) to memory during each instruction cycle.

If we consider programming a simple Von Neumann architecture machine to implement our example FIR filter algorithm, the shortcomings of the architecture become immediately apparent. Even if the processor's data path is capable of completing a multiply-accumulate operation in one instruction cycle, it will take four instruction cycles for the processor to actually perform the multiply-accumulate operation, since the four memory accesses outlined above must proceed sequentially with each memory access taking one instruction cycle. This is one reason why conventional processors often do not perform well on DSP-intensive applications, and why designers of DSP processors have developed a wide range of alternatives to the Von Neumann architecture, which we explore below. Each of these alternatives offers improved memory access bandwidth when compared to the basic Von Neumann architecture. Different processors use very different techniques to achieve this increased bandwidth, and in many cases (mostly in smaller, fixed-point devices) processors place severe restrictions on how this added bandwidth can be used. Such restrictions often contribute significantly to the difficulty of developing high-performance software for DSP processors.

Harvard Architectures

The name *Harvard architecture* refers to a memory structure wherein the processor is connected to two independent memory banks via two independent sets of buses. In the original Har-

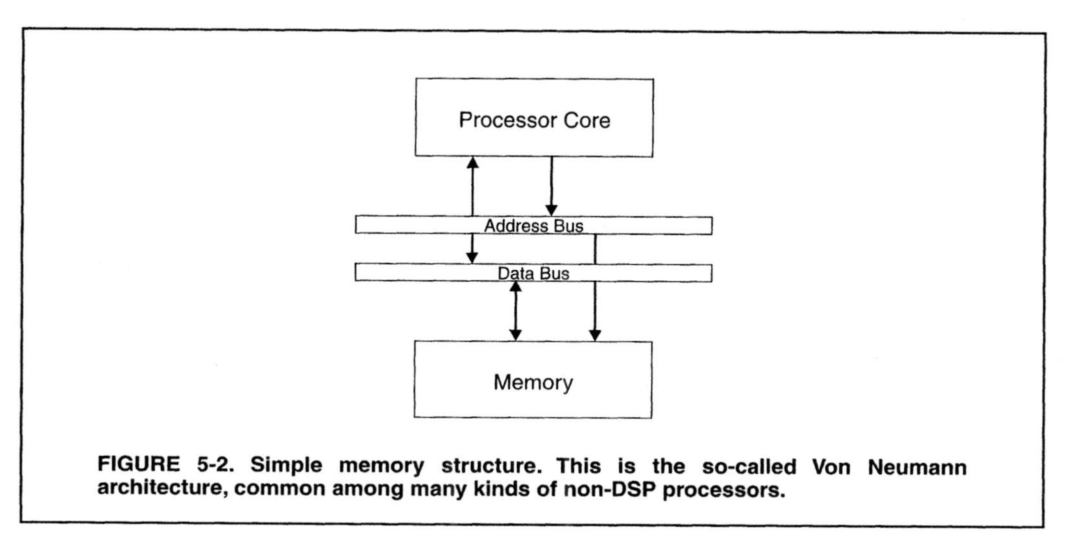

FIGURE 5-2. Simple memory structure. This is the so-called Von Neumann architecture, common among many kinds of non-DSP processors.

vard architecture, one memory bank holds program instructions and the other holds data. Commonly, this concept is extended slightly to allow one bank to hold program instructions *and* data, while the other bank holds data only. This "modified" Harvard architecture is shown in Figure 5-3.

The key advantage of the Harvard architecture is that *two* memory accesses can be made during any one instruction cycle. Thus, the four memory accesses required for our example FIR filter can be completed in two instruction cycles.

This type of memory architecture is used in many DSP processor families, including the Analog Devices ADSP-21xx and the AT&T DSP16xx, although on the DSP16xx, writes to memory always take two instruction cycles, so the full potential of the dual-bank structure is not realized.

If two memory banks are better than one, then one might suspect that three memory banks would be better still. Indeed, this is the approach adopted by several DSP processor manufacturers. The modified Harvard architectures of the PineDSPCore and OakDSPCore from DSP Group provide three memory banks, each with its own set of buses: a program memory bank and two data memory banks, designated X and Y. These three memories allow the processor to make three independent memory accesses per instruction cycle: one program instruction fetch, one X memory data access, and one Y memory data read. Other processors based on a three-bank modified

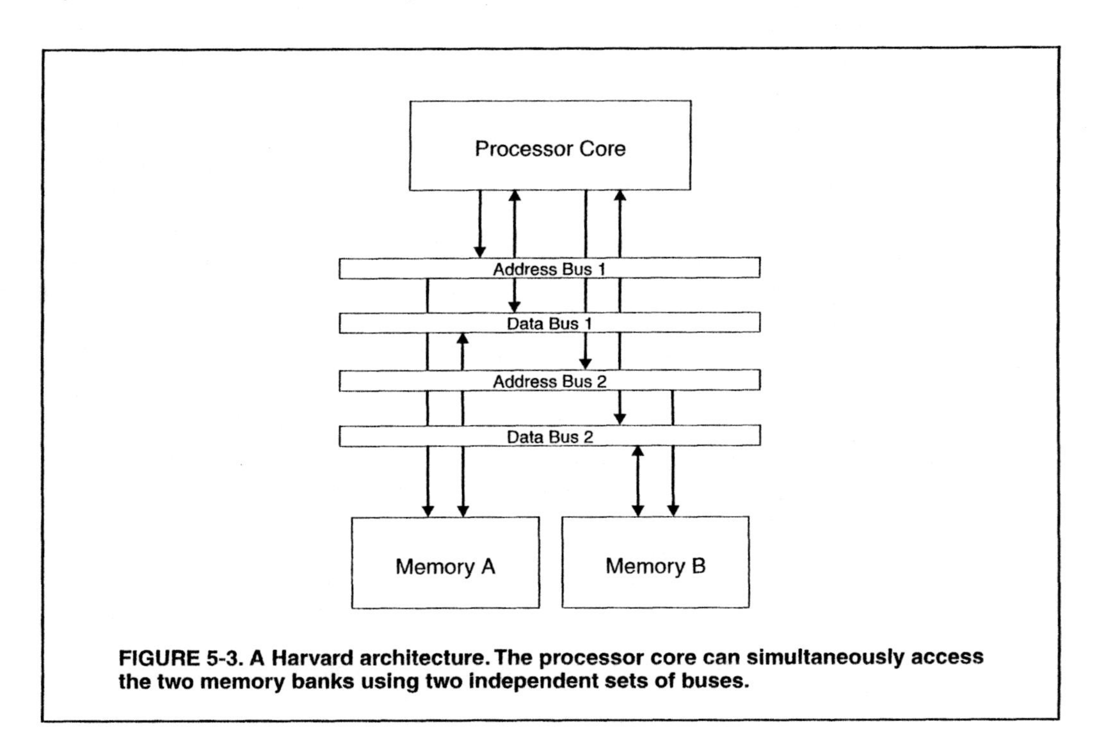

FIGURE 5-3. A Harvard architecture. The processor core can simultaneously access the two memory banks using two independent sets of buses.

Harvard architecture include the Zilog Z893xx, the SGS-Thomson D950-CORE, and the Motorola DSP5600x, DSP563xx, and DSP96002.

For our FIR filter example, recall that we nominally need four memory accesses per instruction cycle in order to compute one filter tap per instruction cycle. Many processors that support only three memory accesses per instruction cycle dispense with the need for a fourth memory access to update the filter delay line by using a technique called *modulo addressing*, which is discussed in Section 5.2.

Because extending multiple memory buses outside the chip is costly, DSP processors generally provide only a single off-chip bus set (i.e., one address and one data bus). Processors with multiple memory banks usually provide a small amount of memory on-chip for each bank. Although the memory banks can usually be extended off-chip, multiple off-chip memory accesses cannot proceed in parallel (due to the lack of a second set of external memory buses). Therefore, if multiple accesses to off-chip memory are requested by an instruction, the instruction execution is extended to allow time for the multiple external accesses to proceed sequentially. Issues relating to external memory are discussed later in this section.

Multiple-Access Memories

As we've just discussed, Harvard architectures achieve multiple memory accesses per instruction cycle by using multiple, independent memory banks connected to the processor data path via independent buses. While a number of DSP processors use this approach, there are also other ways to achieve multiple memory accesses per instruction cycle. These include using fast memories that support multiple, sequential accesses per instruction cycle over a single set of buses, and using multiported memories that allow multiple concurrent memory accesses over two or more independent sets of buses.

Some processors use on-chip memories that can complete an access in one half of an instruction cycle. This means that two independent accesses to a single memory can be completed in sequence. Fast memories can be combined with a Harvard architecture, yielding better performance than could be obtained from either technique alone. For example, consider a modified Harvard architecture with two banks of fast memory. Each bank can complete two sequential memory accesses per instruction cycle. The two banks together can complete four memory accesses per instruction cycle, assuming the memory accesses are arranged so that each memory bank handles two accesses. In general, if the memory accesses cannot be divided in this way so that, for example, three accesses are made to one bank, the processor automatically lengthens the execution of the instruction to allow time for three sequential memory accesses to complete. Thus, there is no risk that a suboptimal arrangement of memory accesses will cause erroneous results; it simply causes the program to run more slowly.

Zoran's ZR3800x combines a modified Harvard architecture with multiple-access memory. This processor combines a single-access program memory bank with a dual-access data memory bank. Thus, one program fetch and two data accesses to on-chip memory can be completed per instruction cycle. The AT&T DSP32xx combines a Von Neumann architecture with multiple access memories. This processor can complete four sequential accesses to its on-chip memory in a single instruction cycle.

Another technique for increasing memory access capacity is the use of *multiported* memories. A multiported memory has multiple independent sets of address and data connections, allowing multiple independent memory accesses to proceed in parallel. The most common type of multiported memory is the *dual-ported* variety, which provides two simultaneous accesses. However, triple- and even quadruple-ported varieties are sometimes used. Multiported memories dispense with the need to arrange data among multiple, independent memory banks to achieve maximum performance. The key disadvantage of multiported memories is that they are much more costly (in terms of chip area) to implement than standard, single-ported memories.

Some DSP processors combine a modified Harvard architecture with the use of multiported memories. The memory architecture shown in Figure 5-4, for example, includes a single-ported program memory with a dual-ported data memory. This arrangement provides one program memory access and two data memory accesses per instruction word and is used in the Motorola DSP561xx processors.

For the most part, the use of fast memories with multiple sequential accesses within an instruction cycle and multiported memories with multiple parallel accesses is limited to what can be squeezed onto a single integrated circuit with the processor core because of limitations on chip input/output performance and capacity. In the case of fast memories, moving the memory (or part of it) off-chip means that significant additional delays are introduced between the processor core

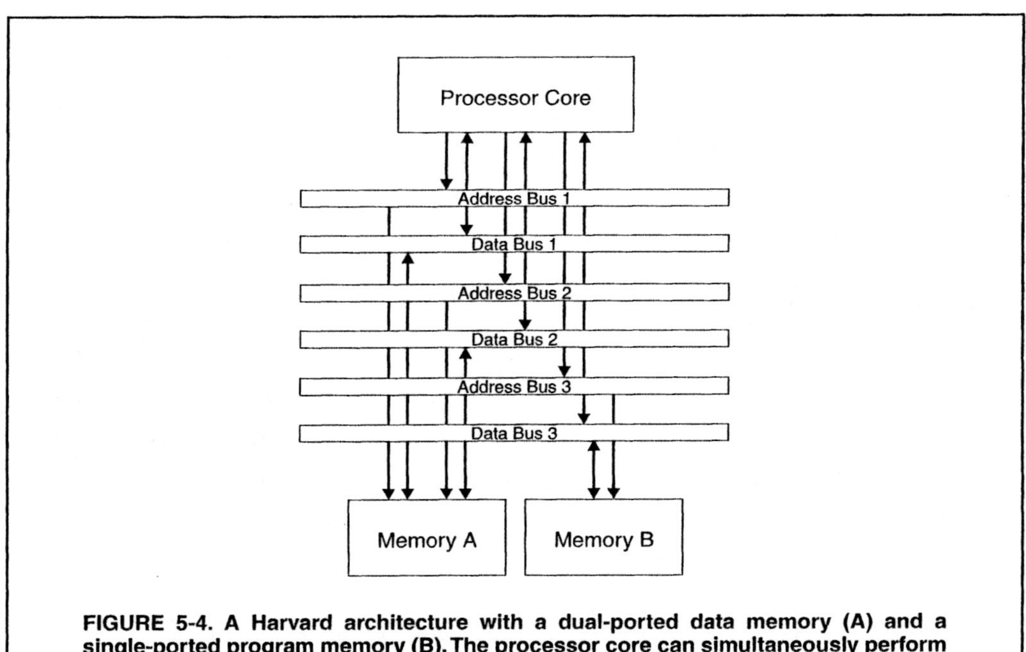

FIGURE 5-4. A Harvard architecture with a dual-ported data memory (A) and a single-ported program memory (B). The processor core can simultaneously perform two accesses to memory bank A and one access to memory bank B using three independent sets of buses.

and the memory. Unless the processor instruction rate is relatively slow, these delays may make it impractical to obtain two or more sequential memory accesses per instruction cycle. In the case of multiported memories, moving all or part of the memory off-chip means that multiple address and data buses must be brought outside the chip. This implies that the chip will need many more I/O pins, which often means that a larger, more expensive package and possibly also a larger die size must be used.

Specialized Memory Write Operations

A few processors provide a specialized mechanism to allow a write to data memory to proceed in parallel with an instruction read and a data read. These processors provide special instructions that allow a parallel write to data memory under certain restricted circumstances. This write operation can be used to shift data along the delay line in an FIR filter implementation. For example, the AT&T DSP16xx normally cannot provide both a data memory write and a data memory read in less than three instruction cycles. However, under certain circumstances, an operand register value can be written to one memory location and then loaded with a value from another memory location (essentially a specialized swap operation) in only two instruction cycles. Texas Instruments' fixed-point DSPs provide a similar kind of operation: a value in memory can be loaded into the operand register and also copied to the next higher location in memory.

5.2 Features for Reducing Memory Access Requirements

Some DSP processors provide special features designed to reduce the number of memory accesses required to perform certain kinds of operations. Under some circumstances these features allow such processors to achieve equal performance to other processors that provide more memory bandwidth. Because a processor with more memory bandwidth is generally more expensive, features that reduce memory access requirements also tend to reduce processor cost. Of course, they may also increase execution time or software development time, and therefore represent a trade-off that must be carefully considered by the system designer.

Program Caches

Some DSP processors incorporate a *program cache*, which is a small memory within the processor core that is used for storing program instructions to eliminate the need to access program memory when fetching certain instructions. Avoiding a program instruction fetch can free a memory access to be used for a data read or write, or it can speed operation by avoiding delays associated with slow external (off-chip) program memory.

DSP processor caches vary significantly in their operation and capacity. They are generally much smaller and simpler than the caches associated with general-purpose microprocessors. We briefly discuss each of the major types of DSP processor caches below.

The simplest type of DSP processor cache is a single-instruction *repeat buffer.* This is a one-word instruction cache that is used with a special *repeat* instruction. A single instruction that is to be executed multiple times is loaded into the buffer upon its first execution; immediately, subsequent executions of the same instruction fetch the instruction from the cache, freeing the program memory to be used for a data read or write access. For example, the Texas Instruments TMS320C2x and TMS320C5x families provide one program memory access and one data memory access per instruction cycle. However, when an instruction is placed in the repeat buffer for repeated execution, the second and subsequent executions of the instruction can perform *two* memory accesses (one to program memory to fetch one data value and one to data memory to fetch another data value). Thus, when the repeat instruction is used, the processor can achieve performance comparable to a processor that provides three memory accesses per instruction cycle. The obvious disadvantage to the repeat buffer approach is that it works on only one instruction at a time, and that instruction must be executed repeatedly. While this is very useful for some algorithms (e.g., dot-product computation), it does not help for algorithms in which a block of multiple instructions must be executed repeatedly as a group.

The repeat buffer concept can be extended to accommodate more than one program instruction. For example, the AT&T DSP16xx provides a 16-entry repeat buffer. The DSP16xx buffer is loaded when the programmer specifies a block of code of 16 or fewer words to be repeated using the *repeat* instruction. The first time through, the block of instructions are read from program memory and copied to the buffer as they are executed. During each repetition, the instructions are read from the buffer, freeing one additional memory access for a data read or write. As with the TMS320C2x and TMS320C5x, the DSP16xx can achieve two data transfers per instruction cycle when the repeat buffer is used. Multiword repeat buffers work well for algorithms that contain loops consisting of a modest number of instructions. This type of loop is quite common in DSP algorithms, since many (if not most) DSP algorithms contain groups of several instructions that are executed repeatedly. Such loops are often used in filtering, transforms, and block data moves.

A generalization of the multi-instruction repeat buffer is a simple *single-sector instruction cache*. This is a cache that stores some number of the most recent instructions that have been executed. If the program flow of control jumps back to an instruction that is in cache (a *cache hit*), the instruction is executed from the cache instead of being loaded from program memory. This frees an additional memory access for a data transfer, and avoids a speed penalty that may be associated with accessing slow off-chip program memory. The limitation on this type of cache is that it can be used to access only a single, contiguous region of program memory. When a program control flow change (for example, a *branch* instruction or an interrupt service routine) accesses a program memory location that is not already contained in the cache, the previous contents of the cache are invalidated and cannot be used.

The difference between the single-sector instruction cache and the multiword repeat buffer is that the cache is loaded with each instruction as it is executed and tracks the addresses of the instructions in the cache. If the program flow of control jumps to a program address that is contained in the cache, the processor detects this and accesses the instructions out of the cache. This means that the cache can be accessed by a variety of instructions, such as *jump*, *return*, and so on. With the repeat buffer, only the repeat instruction can be used to access instructions in the cache. This means that a repeat buffer cannot be used to hold branch instructions. An example of a processor using a single-sector cache is the Zoran ZR3800x. As with multiword repeat buffers, single-sector caches are useful for a wide range of DSP processor operations that involve repetitively executing small groups of instructions.

A more flexible structure is a cache with multiple independent sectors. This type of cache functions like the simple single-sector instruction cache, except that two or more independent segments of program memory can be stored. For example, the cache in the Texas Instruments TMS320C3x contains two sectors of 32 words each. Each sector can be used to store instructions from an independent 32-word region of program memory. If the processor attempts to fetch an instruction from an external memory location that is stored in the cache (a *cache hit*), the external access is not made, and the word is taken from the cache. If the memory location is not in the cache (a *cache miss*), then the instruction is fetched from external memory, and the cache is updated in one of two ways. If the external address was from one of the two 32-word sectors currently associated with the cache, then the word is stored in the cache at the appropriate location within that sector. If the external address does not fall within the two 32-word sectors currently being monitored by the cache, then a *sector miss* occurs. In this case, the entire contents of one of the sectors is discarded and that sector becomes associated with the 32-word region of memory containing the accessed address. In the case of Texas Instruments processors, the algorithm used to determine which cache sector should be discarded when a sector miss occurs is the *least-recently-used* (or LRU) algorithm. This algorithm keeps track of when each cache sector has been accessed. When a cache sector is needed to load new program memory locations, the algorithm selects the cache sector that has not been read from for the longest time.

Some DSP processors with instruction caches provide special instructions or configuration bits that allow the programmer to lock the contents of the cache at some point during program execution or to disable the cache altogether. These features provide a measure of manual control over cache mechanisms, which may allow the programmer to obtain better performance than would be achieved with the built-in cache management logic of the processor. In addition, imposing manual control over cache loading may help software developers to ensure that their code will meet critical real-time constraints.

An interesting approach to caches was introduced by Motorola with the updated DSP96002. This processor allows the internal 1 Kword by 32-bit program memory to be configured either as an instruction cache or as program memory. When the cache is enabled, it is organized into eight 128-word sectors. Each sector can be individually locked and unlocked. Motorola's more recent DSP563xx family includes a similar dual cache/memory construct.

A variation on the multisector caches just discussed is the Analog Devices ADSP-210xx cache. The ADSP-210xx uses a two-bank Harvard architecture; instructions that access data from program memory require two accesses and therefore cause contention for program memory. Because the ADSP-210xx cache is loaded only with instructions whose execution causes contention for program memory access, the cache is more efficient than a traditional cache, which stores every instruction fetched.

Although DSP processor caches are in some cases beginning to approach the sophistication of caches found in high-performance general-purpose processors, there are still some important differences. In particular, DSP processor caches are used only for program instructions, not for data. A cache that accommodates data as well as instructions must include a mechanism for updating both the cache and external memory when a data value held in the cache is modified by the program. This adds significantly to the complexity of the cache hardware.

Modulo Addressing

As we've just discussed, cache memories reduce the number of accesses to a processor's main memory banks required to accomplish certain operations. They do this by acting as an additional, specialized memory bank. In special circumstances, it is possible to use other techniques to reduce the number of total memory accesses (including use of a cache, if one exists) required to accomplish certain operations. One such technique is *modulo addressing*, which is discussed in detail in Chapter 6. Modulo addressing enables a processor to implement a delay line, such as the one used in our FIR filter example, without actually having to move the data values in memory. Instead, data values are written to one memory location and remain there until they are no longer needed. The effect of data shifting along a delay line is simulated by manipulating memory pointers using modulo arithmetic. This technique reduces the number of simultaneous memory accesses required to implement the FIR filter example from four per instruction cycle to three per instruction cycle.

Algorithmic Approaches

Although not a DSP processor feature, another technique for reducing memory access requirements is to use algorithms that exploit data locality to reduce the number of memory accesses needed. DSP algorithms that operate on blocks of input data often fetch the same data from memory multiple times during execution. A clever programmer can reuse previously fetched data to reduce the number of memory accesses required by an algorithm. For example, Figure 5-5 illustrates an FIR filter operating on a block of two input samples. Instead of computing output samples one at a time, the filter instead computes two output samples at a time, allowing it to reuse previously fetched data. This reduces the memory bandwidth required from one instruction fetch and two data fetches per instruction cycle to one instruction fetch and one data fetch per instruction cycle. At the expense of slightly larger code size, this technique allows (for example) FIR filter outputs to be computed at one instruction cycle per tap while requiring less memory bandwidth than a more straightforward approach. This technique is heavily used on IBM's Mwave family of DSP processors, which have limited memory bandwidth. Within IBM the technique is known as the "Zurich Zip," in honor of the researcher at IBM Zurich Laboratories who popularized it.

5.3 Wait States

As the name implies, *wait states* are states in which the processor cannot execute its program because it is waiting for access to memory. Wait states occur for three reasons: contention, slow memory, and bus sharing.

Conflict wait states occur when the processor attempts to make multiple simultaneous accesses to a memory that cannot accommodate multiple accesses. This may occur, for example, when a single bank of single-access memory contains both instruction words and data. Since most DSP processors are heavily pipelined, the execution of a single instruction is often spread across several instruction cycles. Therefore, conflict wait states can arise even when a particular single instruction does not require more accesses to a given memory bank than that memory bank can support, because adjacent instructions may require memory access at the same time. Pipelining is discussed in detail in Chapter 9.

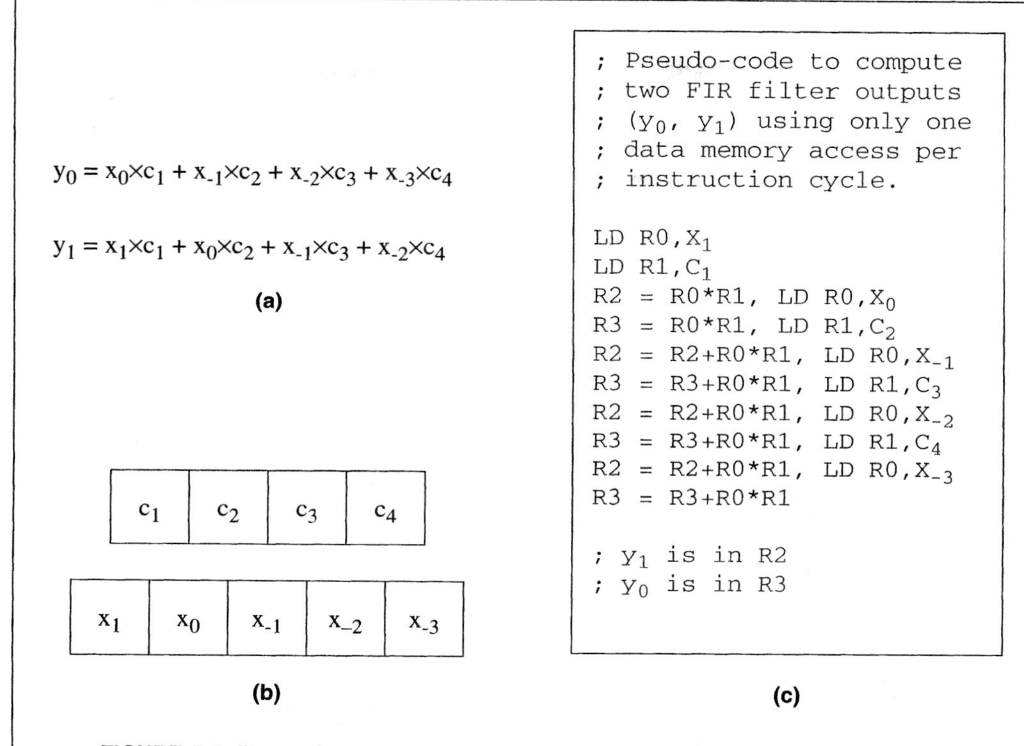

FIGURE 5-5. Illustration of an algorithmic approach to reducing memory access requirements using a block FIR filter with a block size of two samples and four taps. (a) FIR filter equations, (b) memory layout, (c) pseudo-code implementing the FIR filter in essentially one instruction cycle per tap while requiring only one data memory access per instruction cycle.

Almost all processors recognize the need for conflict wait states and automatically insert the minimum number of conflict wait states needed. Exceptions to this are a few members of the AT&T DSP16xx family (the DSP1604, DSP1605, and DSP1616). On these processors, attempting to fetch words from both external program and data memory in one instruction cycle results in a correct program word fetch, but the fetched data word is invalid.

Most DSP processors include one or more small banks of fast on-chip RAM and/or ROM that provide one or more accesses per instruction cycle. In many situations, it is necessary or desirable to expand this memory using off-chip memory that is too slow to support a complete memory access within one processor instruction cycle. Typically this is done to save cost, since slower memory chips are cheaper than faster ones. In these cases, the processor is configured to insert *programmed wait states* during external memory accesses. These wait states are configured by the programmer to deliberately slow down the processor's memory accesses to match the speed of slow memories. Some processors can be programmed to use different numbers of programmed wait states when accessing different regions of off-chip memory, so cost-effective combinations of slower and faster memory can be used.

In some systems, it may not be possible to predict in advance precisely how many wait states will be required to access external memory. For example, when the processor shares an external memory bus with one or more other processors, the processor may have to wait for another processor to relinquish the bus before it can proceed with its own access. Similarly, if dynamic memory (DRAM) is used, the processor may have to wait while the DRAM controller refreshes the DRAM. In these cases, the processor must have the ability to dynamically insert *externally requested wait states* until it receives a signal from an external bus or memory controller that the external memory is ready to complete the access. For example, the Texas Instruments TMS320C5x provides a special READY pin that can be used by external hardware to signal the processor that it must wait before continuing with an external memory access.

The length of a wait state relative to the length of a processor instruction cycle varies from processor to processor. Wait state lengths typically range from one quarter of an instruction cycle (as on the AT&T DSP32C) to a full instruction cycle (as on most processors). Shorter wait states allow more efficient operation, since the delay from the time when the external memory is ready for an access to the time when the wait state ends and the processor begins the access will likely be shorter.

5.4 ROM

DSP processors that are intended for low-cost, embedded applications like consumer electronics and telecommunications equipment provide on-chip read-only memory (ROM) to store the application program and constant data. Some manufacturers offer multiple versions of their processors: a version with internal RAM for prototyping and for low-volume production, and a version with factory-programmed ROM for large-volume production. On-chip ROM sizes typically range from 256 words to 36 Kwords.

Texas Instruments offers versions of some of its processors (e.g., the TMS320P17 and TMS320P25) with one-time-programmable ROM on-chip. These devices can be programmed by the system manufacturer using inexpensive PROM programmers, either for prototyping or for low- or medium-volume production.

For applications requiring more ROM than is provided on-chip by the chosen processor, external ROM can be connected to the processor through its external memory interface. Typically, multiple ROM chips are used to create a bank of memory whose width matches the width of the program word of the processor. However, some processors have the ability to read their initial (boot) program from an inexpensive byte-wide external ROM. These processors construct instruction words of the appropriate width by concatenating bytes from the ROM.

5.5 External Memory Interfaces

DSP processors' external memory interfaces differ in three main features: number of memory ports, sophistication and flexibility of the interface, and timing requirements.

Most DSP processors provide a single external memory port consisting of an address bus, a data bus, and a set of control signals, even though most DSP processors have multiple independent memory banks on-chip. This is because extending buses off-chip requires large numbers of package pins, which increase the cost of the processor. Most processors with multiple on-chip memory banks provide the flexibility to use the external memory port to extend any of the internal memory banks off-chip. However, the lack of multiple external memory ports usually means that multiple accesses cannot be made to external memory locations within a single instruction cycle, and programs attempting to do so will incur a performance penalty. Figure 5-6 illustrates a typical DSP processor external memory interface, with three independent sets of on-chip memory buses sharing one external memory interface.

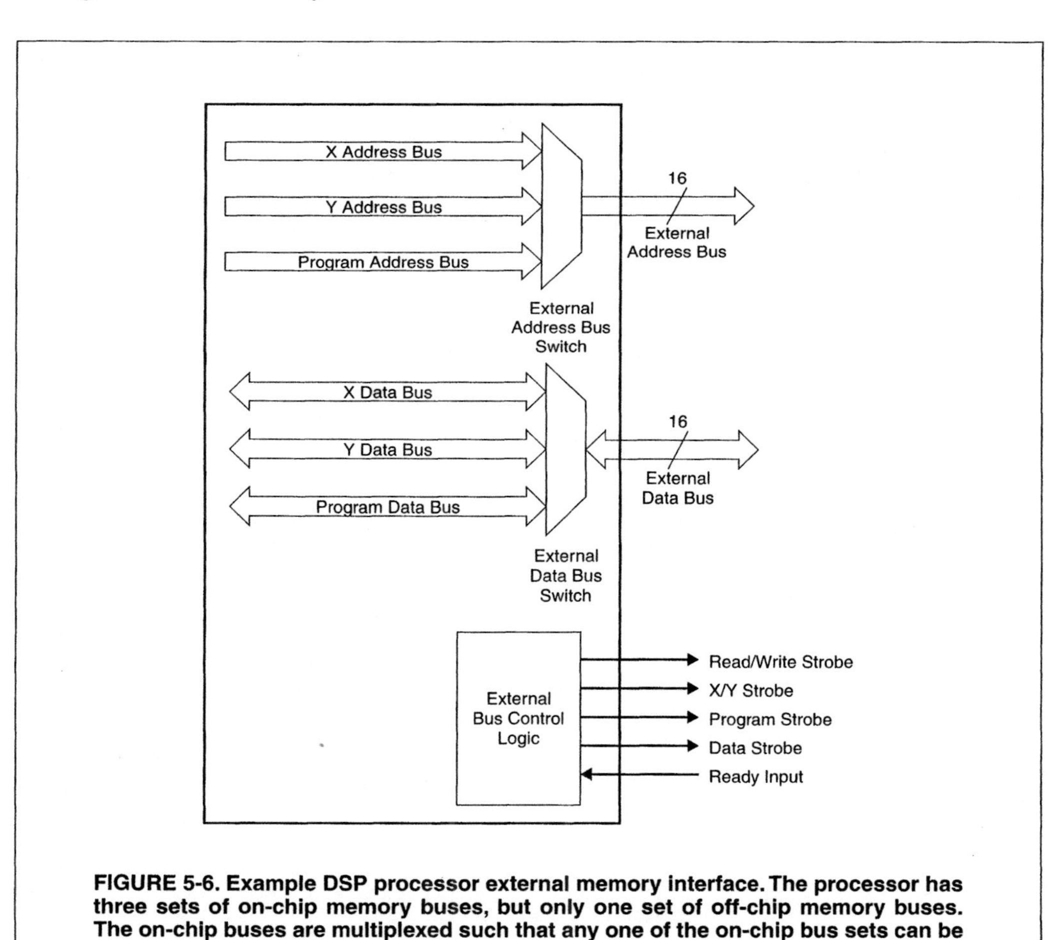

FIGURE 5-6. Example DSP processor external memory interface. The processor has three sets of on-chip memory buses, but only one set of off-chip memory buses. The on-chip buses are multiplexed such that any one of the on-chip bus sets can be connected to the off-chip bus set.

Some DSP processors do provide multiple off-chip memory ports. The Analog Devices ADSP-21020 provides an external program memory port (24-bit address, 48-bit data) and an external data memory port (32-bit address, 32-bit data). The Texas Instruments TMS320C30 provides one 24-bit address, 32-bit data external memory port, and one 13-bit address, 32-bit data external memory port, while the TMS320C40 has two identical 31-bit address, 32-bit data external memory ports. Similarly, the Motorola DSP96002 provides two identical 32-bit address and data bus sets. The cost of these devices is correspondingly higher than that of comparable processors with only one external memory port.

DSP processor external memory interfaces vary quite a bit in flexibility and sophistication. Some are relatively simple and straightforward, with only a handful of control pins. Others are much more complex, providing the flexibility to interface with a wider range of external memory devices and buses without special interfacing hardware. Some of the features distinguishing external memory interfaces are the flexibility and granularity of programmable wait states, the inclusion of a *wait* pin to signal the availability of external memory, bus request and bus grant pins (discussed below), and support for page-mode DRAM (discussed below).

High-performance applications must often use fast static RAM devices for off-chip memory. In such situations, it is important for system hardware designers to scrutinize the timing specifications for DSP processors' external memory ports. Because timing specifications can vary significantly among processors, it is common to find two processors that have the same instruction cycle time but have very different timing specifications for off-chip memory. These differences can have a serious impact on system cost, because faster memories are significantly more expensive than slower memories. Hardware design flexibility is also affected, since more stringent timing specifications may constrain the hardware designer in terms of how the interface circuitry is designed and physically laid out.

Manual Caching

Whether or not a processor contains a cache, it is often possible for software developers to improve performance by explicitly copying sections of program code from slower or more congested (in terms of accesses) memory to faster or less congested memory. For example, if a section of often-used program code is stored in a slow, off-chip ROM, then it may make sense to copy that code to faster on-chip RAM, either at system start-up or when that particular program section is needed.

Multiprocessor Support in External Memory Interfaces

DSP processors intended for use in multiprocessor systems often provide special features in their external memory interfaces to simplify the design and enhance the performance of such systems. The first and most obvious of these features is the provision of two external memory ports, mentioned above. The availability of two external memory ports means that one port can be connected to a local, private memory, while the other is connected to a memory shared with other processors. For example, the Motorola DSP96002 includes two external memory ports expressly for use in such multiprocessor configurations.

When a multiprocessor system includes two or more processors that share a single external memory bus, a mechanism must be provided for the processors to negotiate control of the bus

(*bus arbitration*) and to prevent the processors that do not have control of the bus from trying to assert values onto the bus. Several DSP processors provide features to facilitate this kind of arrangement, though there are significant differences in the sophistication and flexibility of the features provided. In some cases, a shared bus multiprocessor can be created simply by connecting together the appropriate pins of the processors without the need for any special software or hardware to manage bus arbitration. In other cases, extra software on one or more of the DSP processors and/or external bus arbitration hardware may be required.

An example of basic support for shared bus systems is provided by the Motorola DSP5600x. Two of the DSP processor's pins can be configured to act as *bus request* and *bus grant* signals. When an external bus arbitrator (either another processor or dedicated hardware) wants a particular DSP processor to relinquish the shared bus, it asserts that processor's bus request input. The processor then completes any external memory access in progress and relinquishes the bus, acknowledging with the bus grant signal that it has done so. The DSP processor can continue to execute its program as long as no access to the shared bus is required. If an access to the shared bus is required, the processor waits until the bus request signal has been deasserted, indicating that it can again use the shared bus.

The Texas Instruments TMS320C5x provides several features that support multiprocessing. In addition to providing the equivalent of bus request and bus grant signals (called $\overline{\text{HOLD}}$ and $\overline{\text{HOLDA}}$ on the TMS320C5x), the processor also allows an external device to access its *on-chip* memory. To accomplish this, the external device first asserts the TMS320C5x's $\overline{\text{HOLD}}$ input. When the processor responds by asserting $\overline{\text{HOLDA}}$, the external device asserts $\overline{\text{BR}}$, indicating that it wishes to access the TMS320C5x's on-chip memory. The TMS320C5x responds by asserting $\overline{\text{IAQ}}$. The external device can then read and write the TMS320C5x's on-chip memory by driving TMS320C5x's address, data, and read/write lines. When finished, the external device deasserts $\overline{\text{HOLD}}$ and $\overline{\text{BR}}$. This allows the creation of multiprocessor systems that do not require shared memory for interprocessor communications.

A processor feature that simplifies the use of shared variables in shared memory is *bus locking*, which allows a processor to read the value of a variable from memory, modify it, and write the new value back to memory, while ensuring that this sequence of operations is not interrupted by another processor attempting to update the variable's value. This is sometimes referred to as an *atomic test-and-set* operation. The Texas Instruments TMS320C3x and TMS320C4x processors provide special instructions and hardware support for bus locking. Texas Instruments refers to these operations as "interlocked operations."

The Analog Devices ADSP-2106x offers a sophisticated shared bus interface. The processor provides on-chip bus arbitration logic that allows direct interconnection of up to six ADSP-2106x devices with no special software or external hardware required for bus arbitration. In addition, the processor allows one DSP processor in a shared-bus configuration to access another processor's on-chip memory, much like on the Texas Instruments TMS320C5x family. This means that interprocessor data moves will not necessarily have to transit through an external shared memory.

In addition to special external memory interface features, the Analog Devices ADSP-2106x and the Texas Instruments TMS320C4x families provide special communications ports to facilitate connections within multiprocessor systems. Features of this type are discussed in detail in Chapter 10.

Dynamic Memory

All of the writable memory found on DSP processors and most of the memory found in systems based on DSP processors is *static* memory, also called SRAM (for static random-access memory; a better name would have been static read and write memory). Static memory is simpler to use and faster than *dynamic* memory (DRAM), but it also requires more silicon area and is more costly for a given number of bits of memory. The key operational attribute distinguishing static from dynamic memories is that static memories retain their data as long as power is available. Dynamic memories must be *refreshed* periodically; that is, a special sequence of signals must be applied to reinforce the stored data, or it eventually (typically in a few tens of milliseconds) is lost. In addition, interfacing to static memories is usually simpler than interfacing to dynamic memories; the use of dynamic memories usually requires a separate, external DRAM controller to generate the necessary control signals.

Because of the increasing proliferation of DSP processors into low-cost, high-volume products like answering machines and personal computer add-in cards, there has been increased interest in using dynamic memory in DSP systems. DRAM can also be attractive for systems that require large quantities of memory, such as large-scale multiprocessor systems.

One way to get faster, static RAM-like performance from slower, dynamic RAM is the use of *paged* or *static column* DRAM. These are special types of DRAM chips that allow faster than normal access when a group of memory accesses occurs within the same region (or *page*) of memory. Some DSP processors, including the Motorola DSP96002, the Analog Devices ADSP-210xx, and the Texas Instruments TMS320C3x and TMS320C4x provide memory page boundary detection capabilities. These capabilities generally consist of a set of programmable registers, which the programmer uses to specify the locations of page boundaries in external memory, and circuitry to detect when external memory accesses cross page boundaries. In most cases, when the processor detects that a memory access has crossed a page boundary, it asserts a special output pin. It is then up to the external DRAM controller to use a processor input pin to signal back to the processor that it must delay its access by inserting wait states while the controller readies the DRAM for access to a new page.

As mentioned above, the use of DRAM as external memory for a DSP processor usually requires the use of an external DRAM controller chip. This additional chip may increase the manufacturing cost of the design, which partly defeats the reason for using DRAM in the first place. To address this problem, some DSP processors now incorporate a DRAM controller on-chip. The Motorola DSP56004 and DSP56007, for example, provide on-chip DRAM interfaces that include support for page-mode DRAM.

Direct Memory Access

Direct memory access (DMA) is a technique whereby data can be transferred to or from the processor's memory without the involvement of the processor itself. DMA is typically used to provide improved performance for input/output devices. Rather than have the processor read data from an I/O device and copy the data into memory or vice versa, a separate *DMA controller* can handle such transfers more efficiently. This DMA controller may be a peripheral on the DSP chip itself or it may be implemented using external hardware.

Any processor that has the simple bus request/bus grant mechanism described above can be used with an external DMA controller that accesses external memory. Typically, the processor loads the DMA controller with control information including the starting memory address for the transfer, the number of data words to be transferred, the direction of the transfer, and the source or destination peripheral. The DMA controller uses the bus request pin to notify the DSP processor that it is ready to make a transfer to or from external memory. The DSP processor completes its current instruction, relinquishes control of external memory, and signals the DMA controller via the bus grant pin that the DMA transfer can proceed. The DMA controller then transfers the specified number of data words and optionally signals completion to the processor through an interrupt.

Some more sophisticated DSP processors include a DMA controller on-chip that can access internal and external memory. These DMA controllers vary in their performance and flexibility. In some cases, the processor's available memory bandwidth may be large enough to allow DMA transfers to occur in parallel with normal program instruction and data transfers without any impact on performance. For example, the Texas Instruments TMS320C4x contains a DMA controller that, combined with the TMS320C4x's on-chip memory and on-chip DMA address and data buses, can complete one memory access per instruction cycle independent of the processor. The Motorola DSP96002, the Texas Instruments TMS320C3x family, and the Analog Devices ADSP-2106x family all include on-chip DMA controllers with similar capabilities.

Some DMA controllers can manage multiple DMA transfers in parallel. Such a DMA controller is said to have multiple *channels*, each of which can manage one transfer, and each of which has its own set of control registers. The TMS320C4x DMA controller supports six channels, the Analog Devices ADSP-2106x supports ten channels, and the Motorola DSP96002 can handle two channels. Each channel can be used for memory-memory or memory-peripheral transfers.

In contrast, the AT&T DSP3210 includes a more limited, two-channel DMA controller that can only be used for transfers to and from the processor's internal serial port. Since the DSP3210 does not have extra memory bandwidth, the currently executing instruction is forced to wait one cycle when the DMA controller accesses memory. This arrangement (where the processor is suspended during DMA bus accesses) is called *cycle stealing*. The Analog Devices ADSP-21xx provides a similar capability through a mechanism that Analog Devices calls *autobuffering*.

5.6 Customization

We've already mentioned that many DSP processor vendors offer versions of their processors that are customized by placing user-specified programs and/or data into the on-chip ROM. In addition, several vendors can produce DSP core-based ASICs or customizable DSPs (see Chapter 4), which provide the user with more flexibility. These approaches may allow the user to specify memory sizes and configurations (for example, the mix of ROM and RAM) that are best suited to the application at hand. DSP processor vendors offering customizable DSPs or DSP core-based ASICs include AT&T, Clarkspur Design, DSP Group, SGS-Thomson, Tensleep Design, Texas Instruments, and several other vendors.

Chapter 6

Addressing

As we explained in Chapter 4 on data paths, operands used by different functional units of the data path may come from various sources within the processor and external to it. *Operands* here encompass both input data that is processed within an instruction as well as the output data that results from the instruction's execution. *Addressing* refers to the means by which the locations of operands are specified for instructions in general.

Many types of addressing (often called *addressing modes*) exist; for example, operands can be obtained directly from a register or directly from an internal or external memory address. Other addressing modes may seem arcane to the uninitiated but are often useful. In this chapter, we review the types of addressing available on DSP processors, emphasizing the types that are most important for DSP applications.

Most processors support a subset of the addressing modes described here, with varying flexibility and performance levels. Further, they may place restrictions on the use of a given mode. In many cases, a particular mode can be used only with a subset of the processor's instruction set. In more extreme cases, a specific addressing mode may be available with only one particular instruction. It is also common for a single instruction to combine two or more addressing modes, using different modes for different operands. Motivations for restricting support of addressing modes are related to instruction set orthogonality, which is discussed in Chapter 7.

Most DSP processors include one or more special *address generation units* (AGUs) that are dedicated to calculating addresses. Manufacturers refer to these units by various names. For example, Analog Devices calls their AGU a *data address generator*, and AT&T calls theirs a *control arithmetic unit*. An AGU can perform one or more complex address calculations per instruction cycle without using the processor's main data path. This allows address calculations to take place in parallel with arithmetic operations on data, improving processor performance. The differences among address generation units are manifested in the types of addressing modes provided and the capability and flexibility of each addressing mode. Addressing modes are discussed in detail below.

6.1 Implied Addressing

Implied addressing means that the operand addresses are implied by the instruction; there is no choice of operand locations. In these cases, the operands are in registers. For example, in the AT&T DSP16xx, all multiplication operations take their inputs from the multiplier input registers X and Y, and deposit their result into the product register P. The fact that the programmer has selected a multiplication operation *implies* the location of the operands. Thus, although AT&T's assembly language syntax for a multiplication operation is

```
P = X * Y
```

it could just as easily be written simply as

```
*
```

because the programmer does not get a choice of operand registers. AT&T's syntax makes sense, though, because it makes the implied choice of registers explicit and thereby makes the assembly code easier to understand.

6.2 Immediate Data

With immediate data, the operand itself (as opposed to the location where the operand is stored) is encoded in the instruction word or in a separate data word that follows the instruction word. In Analog Devices' ADSP-21xx assembly language syntax, for example, immediate addressing can be used to load a constant into a register, as in

```
AX0 = 1234
```

which causes the constant 1234 to be loaded into the register AX0. Note that in this example the constant value is considered to be specified as immediate data; the destination register, AX0, is specified using *register-direct* addressing (see below).

With immediate addressing, small data words (typically half the size of the instruction word or less) can usually be encoded in the instruction word specifying the operation, resulting in a single-word instruction (see the discussion of short addressing modes below). Larger data words have to be placed in a separate memory word from the instruction. This requires the processor to read two words from program memory before the instruction can be executed, slowing program execution.

6.3 Memory-Direct Addressing

With memory-direct addressing (sometimes called *absolute addressing*), the data being addressed reside in memory at an address encoded in the instruction word or in a separate data word following the instruction word. For example, using Analog Devices' ADSP-21xx assembly language,

```
AX0 = DM(1000)
```

causes the data located at address 1000 to be loaded into the register AX0.

As with immediate data, small addresses can be encoded in the instruction word specifying the operation, resulting in a single-word instruction. Larger addresses have to be placed in a memory word separate from the instruction. This requires the processor to read two words from program memory before the instruction can be executed, slowing program execution.

6.4 Register-Direct Addressing

With register-direct addressing, the data being addressed reside in a register. The programmer specifies the register as part of the instruction. For example, using Texas Instruments' TMS320C3x assembly language, the instruction

```
SUBF R1, R2
```

subtracts the value in register R1 from the value in register R2 and places the result back in register R2. The distinction between register-direct addressing and implied addressing (discussed above) is that register-direct addressing specifies a choice of registers, whereas with implied addressing the nature of the instruction itself specifies the registers. For example, the TMS320C3x subtract instruction just shown can use any two of eight extended-precision operand registers on the chip.

Register-direct addressing is very important in processors where operands for arithmetic operations always (or mostly) come from registers rather than from memory.

6.5 Register-Indirect Addressing

With register-indirect addressing, the data being addressed reside in memory, and the address of the memory location containing the data is held in a register, as shown in Figure 6-1. Some processors have a special group of registers (called *address registers*) reserved for holding

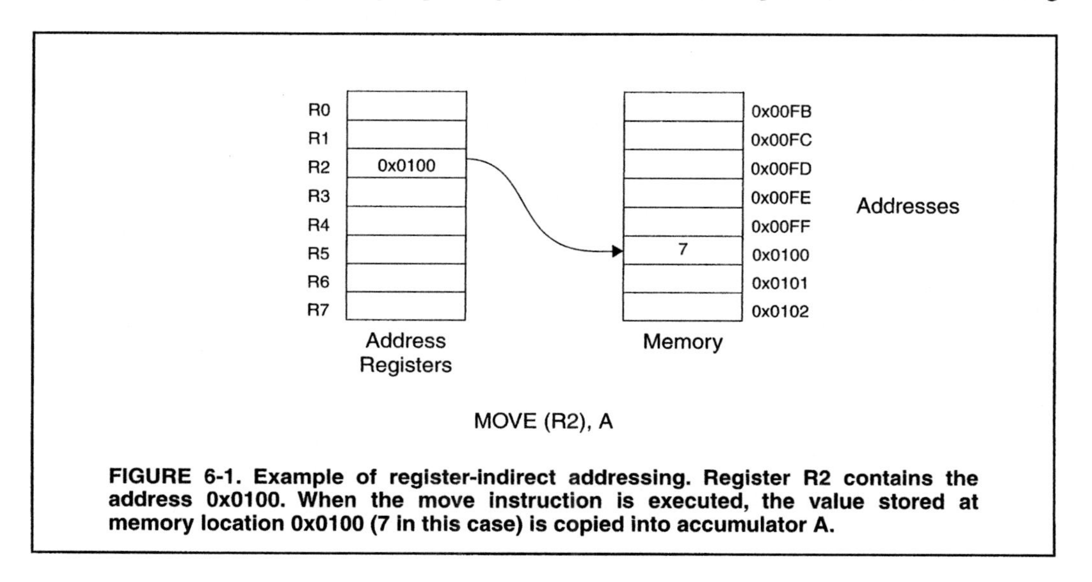

FIGURE 6-1. Example of register-indirect addressing. Register R2 contains the address 0x0100. When the move instruction is executed, the value stored at memory location 0x0100 (7 in this case) is copied into accumulator A.

addresses; other processors have general-purpose registers that can store addresses or data. Register-indirect addressing is, for two reasons, one of the most important addressing modes found on most DSP processors. First, register-indirect addressing lends itself naturally to working with arrays of data, which are common in DSP applications. This point is expanded upon below in the section on register-indirect addressing with pre- and post-increment. Second, register-indirect addressing is efficient from an instruction-set point of view: it allows powerful, flexible addressing while requiring relatively few bits in an instruction word.

As an example of register-indirect addressing, consider the AT&T DSP32xx instruction

```
A0 = A0 + *R5
```

which causes the value stored in the memory location pointed to by the contents of register R5 to be added to the value in the accumulator register A0. The result is stored in the accumulator register A0. In this instruction, the input operand "*R5" is specified using register-indirect addressing: R5 is used as an address register; it holds the address of the memory location that contains the actual operand.

In AT&T's assembly language, the "*" symbol, when used in this manner, is interpreted to mean "the contents of the memory location pointed to by...". This meaning is familiar to programmers who use the C programming language. Other processors have different syntaxes for register-indirect addressing; many use "(Rn)" to indicate the memory location pointed to by address register Rn.

There are many variations on register-indirect addressing; these variations are described next.

Register-Indirect Addressing with Pre- or Post-Increment

Because many DSP applications involve accessing blocks of data in memory, DSP processors include a number of addressing modes designed to speed access to groups of related memory locations. The class of addressing modes we call *register-indirect addressing with pre- or post-increment* starts with the basic register-indirect addressing capability described above, but adds operations that change the value(s) in the address register(s) being used, for example, by adding an increment of one to the value in an address register. The modification can occur before the address is used to access data in memory *(pre-increment)* or after the address is used to access memory *(post-increment)*.

The most common kind of modification applied to address registers is to add an increment of one after the address register is used to access the operand. This is called *post-increment by one*. This addressing mode provides a simple means for stepping through an array of values stored sequentially in memory. For example, the DSP32xx instruction

```
A0 = A0 + *R5++
```

adds the value in the memory location pointed to by register R5 to the accumulator A0 and *then* increments the value in the address register R5 so it points to the next memory location.

Similarly, the instruction

```
A0 = A0 + *R5--
```

causes the address register R5 to be decremented to point to the previous memory location after the addition operation is performed.

Some processors also provide the ability to add or subtract values other than one to or from an address register. For example, the DSP32xx instruction

```
A0 = A0 + *R5++R17
```

adds the value in the memory location pointed to by register R5 to the accumulator and *then* increases the value in the address register R5 by the value stored in register R17. Since the value in register R17 can be negative, this same instruction format can be used to decrease the value stored in an address register. In some processors, a special group of registers is provided to hold pre- or post-increment values. In other processors, the increment value can be stored in a general-purpose register. The register that holds the increment value (R17 in this example) is called an *offset register.* (Note that some vendors use different terminology. For example, Analog Devices calls these registers *modifier registers.*)

An example of a pre-decrement addressing mode is provided by the Motorola DSP5600x:

```
MOVE X:-(R0), A1
```

which decrements address register R0 *before* it is used to point to the memory location containing the data value to be moved into the accumulator register A1. Pre-incrementing usually requires an extra instruction cycle to update the address register contents before they are used to address memory.

Register-Indirect Addressing with Indexing

With indexed addressing, values stored in two address registers (or an address register and a constant) are added together to form an *effective address,* that is, the address that will be used to access memory. The difference between indirect addressing with indexing and indirect addressing with premodification (discussed above) is that with indexing, the values held in the address registers are not modified. For example, the DSP5600x instruction

```
MOVE Y1, X:(R6 + N6)
```

causes the value in register Y1 to be moved to the memory location in the X memory space whose address is the sum of the values in registers R6 and N6. Neither R6 nor N6 is modified. The index is stored in register N6, which we call an *index register.* (Note that some processor vendors use the term "indexed addressing" to refer to register-indirect addressing with pre- or post-increment by a value other than one. On these processors, what we would call "offset" registers are sometimes called "index" registers.)

On some processors the index value can be supplied as an immediate value that is encoded in the instruction. For example, the Texas Instruments TMS320C3x instruction

```
LDI *-AR1(1), R7
```

computes an effective address by subtracting one from the value in the address register AR1 and then moves the data in the memory location pointed to by the effective address into register R7. The contents of the address register AR1 are not modified.

Indexed addressing is useful when the same program code is used on multiple sets of data. The index register can be used to point to the beginning of a data set in memory, and the regular address registers can be pre- or postmodified to step through the data in the desired fashion. When the program is ready to operate on the next data set, only the value in the index register needs to be changed.

Indexed addressing is also useful in compilers for communicating arguments to subroutines by passing data on the stack. Compilers commonly dedicate an address register to be used as a *stack frame pointer.* Each time a subroutine is called, the stack frame pointer is set up so that the address pointed to by the stack frame pointer contains the address of the previous stack frame. The next address contains the number of arguments being passed to the subroutine, and subsequent addresses contain the arguments themselves. This way, subroutines can access their arguments in a simple, consistent way. For example, if address register AR1 is designated for use as the stack frame pointer, then in the TMS320C3x, the instruction

```
LDI *+AR1(2), R0
```

could be used by any subroutine to copy its first argument into register R0. The subroutine itself is insulated from knowledge of the exact memory locations of its arguments; all it needs to know is where its arguments are located relative to the current stack frame pointer.

Register-Indirect Addressing with Modulo Address Arithmetic

In this section we introduce the *modulo addressing* mode through an example of its most common application, *circular buffer management.* Many DSP applications need to manage *data buffers.* A data buffer is a section of memory that is used to store data that arrive from an off-chip source or from a previous computation until the processor is ready to process the data. In real-time systems where dynamic memory allocation is prohibitively expensive, the programmer usually must determine the maximum amount of data that a given buffer will need to hold and then set aside a portion of memory for that buffer. These buffers generally use a first-in, first-out (FIFO) protocol, meaning that data values are read out of the buffer in the order in which they arrived.

In managing the movement of data into and out of the buffer, the programmer maintains two pointers, which are stored in registers or in memory: a read pointer and a write pointer. The read pointer points to (that is, contains the address of) the memory location containing the next data value that will be read from the buffer. The write pointer points to the location where the next data value to arrive will be written, as illustrated in Figure 6-2(a). Each time a read or write operation is performed, the read or write pointer is advanced and the programmer must check to see whether the pointer has reached the last location in the buffer. When the pointer reaches the end of the buffer, it is reset to point to the first location in the buffer. The action of checking after each buffer operation whether the pointer has reached the end of the buffer, and resetting it if it has, is time consuming. For systems that make extensive use of buffers, this can cause a significant performance bottleneck.

To address this bottleneck, many DSPs have a special addressing capability that allows them to automatically perform the action of checking after each buffer address calculation whether the pointer has reached the end of the buffer and adjusting it relative to the buffer start location if necessary. This capability is called modulo addressing or *circular* addressing.

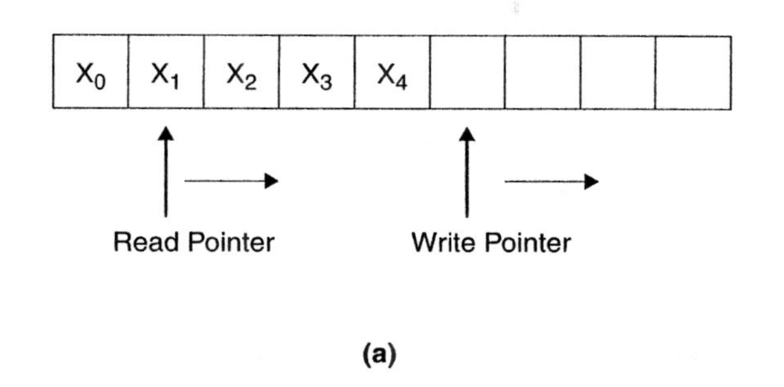

(a)

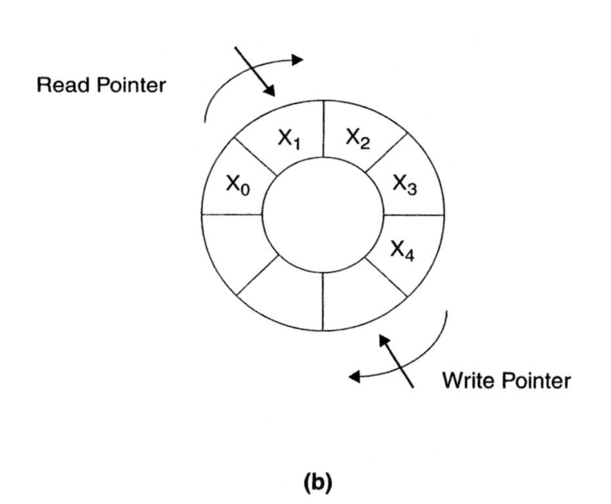

(b)

FIGURE 6-2. (a) A FIFO buffer with linear addressing. Five items of data X_n have arrived in the order X_0, X_1, X_2, X_3, X_4 and have been written into the buffer. Only the first data item, X_0, has been read out of the buffer. After each read or write operation, the corresponding pointer moves to the right. Once either pointer advances to the end of the buffer, it must be reset to point to the beginning of the buffer. (b) The same data in a FIFO buffer with circular addressing. After the read pointer or the write pointer reaches the end of the buffer, it automatically advances to the start of the buffer, making the buffer appear circular to the programmer.

The term *modulo* refers to modulo arithmetic, wherein numbers are limited to a specific range. This is similar to the arithmetic used in a clock, which is based on a 12-hour cycle. When the result of a calculation exceeds the maximum value, it is adjusted by repeatedly subtracting from it the maximum representable value until the result lies within the specified range. For example, 4 hours after 10 o'clock is 2 o'clock (14 modulo 12).

When modulo address arithmetic is in effect, read and write pointers (address registers) are updated using pre- and/or post-increment register-indirect addressing. The processor's address generation unit performs modulo arithmetic when new address values are computed, creating the appearance of a circular memory layout, as illustrated in Figure 6-2(b). Modulo address arithmetic eliminates the need for the programmer to check the read and write pointers to see whether they have reached the end of the buffer and to reset the pointers to the beginning of the buffer once they have reached the end. This results in much faster buffer operations and makes modulo addressing a valuable capability for many applications.

Most recently designed DSP processors provide some support for modulo address arithmetic. However, the depth of this support and the mechanisms used to control it vary from processor to processor. Modulo addressing approaches are discussed in the paragraphs below.

The programmer typically controls modulo addressing in one of two ways. In the first method, the length of the circular buffer is loaded into a special register, often called a *modifier* or *modulo* register. A processor may have only one modifier register, or it may have several. Each modifier register is associated with one or more address registers; whenever a modifier register is loaded with a circular buffer length, its associated address registers automatically use modulo address arithmetic. Because the modifier register contains only the length of the buffer and not its starting address, the modulo addressing circuitry must make some assumptions about the starting address of circular buffers. Typically, circular buffers must start on k-word boundaries, where k is the smallest power of 2 that is equal to or greater than the size of the circular buffer. For example, a 48-word circular buffer would typically have to reside on a 64-word boundary, since 64 is the smallest power of 2 that is equal to or greater than 48. If we imagine dividing a processor's address space into k-word blocks, starting at address 0, then a k-word boundary occurs between every pair of k-word blocks. For example, under this scheme a circular buffer of length 256 could start at address 0, 256, 512, or any other address that is a multiple of 256. Processors implementing this form of modulo addressing (or one like it) include the Texas Instruments TMS320C3x and TMS320C4x, and processors from Motorola, Analog Devices, NEC, and DSP Group.

An alternative approach to modulo addressing uses *start* and *end* registers to hold the start and end addresses of each circular buffer. On some processors, modulo address arithmetic is then used on *any* address registers that point into the region of memory bounded by the addresses in the start and end registers. On other processors, only one address register can be associated with a given circular buffer. Processors using the start/end register approach include the AT&T DSP16xx and the Texas Instruments TMS320C5x.

As suggested above, different processors support different numbers of simultaneously active circular buffers with modulo addressing. For example, the AT&T DSP16xx supports modulo address arithmetic on only one circular buffer at a time, since it has only one start and one end register. The Texas Instruments TMS320C5x supports two, the Motorola DSP561xx supports four, and the Motorola DSP5600x and all Analog Devices processors support eight buffers.

Register-Indirect Addressing with Bit Reversal

Perhaps the most unusual of addressing modes, *bit-reversed* addressing is used only in very specialized circumstances. Some DSP applications make heavy use of the *fast Fourier transform* (FFT) algorithm. The FFT is a fast algorithm for transforming a time-domain signal into its frequency-domain representation and vice versa. However, the FFT has the disadvantage that it either takes its input or leaves its output in a scrambled order. This dictates that the data be rearranged to or from natural order at some point.

The scrambling required depends on the particular variation of the FFT. The radix-2 implementation of an FFT, a very common form, requires reordering of a particularly simple nature, called *bit-reversed* ordering. The term bit-reversed refers to the observation that if the output values from a binary counter are written in reverse order (that is, least significant bit first), the resulting sequence of counter output values will match the scrambled sequence of the FFT output data. This phenomenon is illustrated in Figure 6-3.

Because the FFT is an important algorithm in many DSP applications, many DSP processors include special hardware in their address generation units to facilitate generating bit-reversed address sequences for unscrambling FFT results. For example, the Analog Devices ADSP-210xx provides a bit-reverse mode, which is enabled by setting a bit in a control register. When the processor is in bit-reverse mode, the output of *one* of its address registers is bit-reversed before being applied to the memory address bus.

An alternative approach to implementing bit-reversed addressing is the use of *reverse-carry arithmetic*. With reverse-carry arithmetic, the address generation unit reverses the direction in which carry bits propagate when an increment is added to the value in an address register. If reverse-carry arithmetic is enabled in the AGU, and the programmer supplies the base address and increment value in bit-reversed order, then the resulting addresses will be in bit-reversed order. Reverse-carry arithmetic is provided in the AT&T DSP32C and DSP32xx, for example.

6.6 Short Addressing Modes

As we've illustrated with several examples earlier in this chapter, a shortcoming of several addressing modes is that in general the instruction and the address together require two program words for storage, which increases program size and slows execution. Some processors offer short versions of some of their addressing modes. These short versions squeeze the instruction and the address into a single instruction word, but do so at the expense of some restrictions on the addresses.

Short Immediate Data

With immediate data, the operand itself (as opposed to the location where the operand is stored) is encoded in the instruction word or in a separate data word that follows the instruction

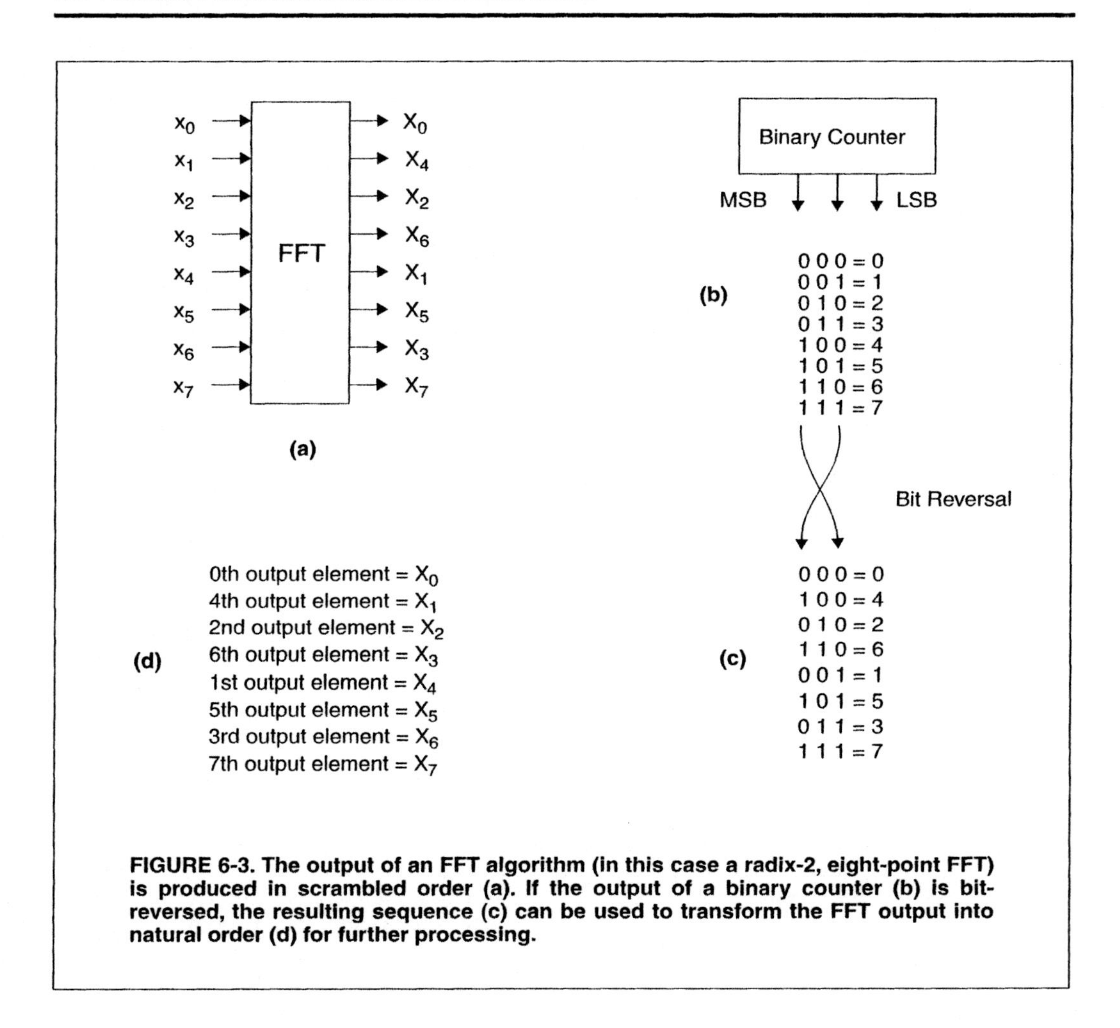

FIGURE 6-3. The output of an FFT algorithm (in this case a radix-2, eight-point FFT) is produced in scrambled order (a). If the output of a binary counter (b) is bit-reversed, the resulting sequence (c) can be used to transform the FFT output into natural order (d) for further processing.

word. In Motorola's DSP5600x assembly language syntax, for example, immediate addressing can be used to load a constant into a register, as in

```
MOVE #1234, A
```

which loads the constant value 1234 into register A. In the general case, this instruction requires two words of program memory: one to store the move instruction and one to store the immediate data themselves. However, if the immediate data are small, it may be possible to fit the instruction and the data into a single program memory word. In the case of the Motorola DSP5600x, immediate data instructions of the form above can be encoded in one program memory word if the data are 12 bits long or smaller.

Short Memory-Direct Addressing

Recall from our earlier discussion of memory-direct addressing that in this addressing mode, the data being addressed reside in memory, and the memory address where those data reside is encoded in the instruction word or in a separate data word following the instruction word. For example, using Motorola's DSP5600x assembly language,

```
MOVE $1000, A
```

causes the data located at address 1000 to be loaded into register A. In this Motorola processor, short memory-direct addressing allows an address of up to six bits in length to be encoded in the same program memory word as the move instruction. This saves one program memory word (and one instruction cycle) compared to the more general version of memory-direct addressing.

In some cases, processors may provide additional versions of short memory-direct addressing that add an implied offset to the short address. For example, the DSP5600x provides a special short memory-direct addressing mode for addressing its I/O registers, which are mapped into the highest 64 locations of its data memory.

Paged Memory-Direct Addressing

With *paged memory-direct addressing*, a special register stores the number of a page (section) of memory that is being addressed. Then, an individual instruction can directly address a word within the page by specifying the relative location of the desired word within the page. To address a word outside of the page, the page register must be updated (or another page register can be used if multiple page registers are provided).

For example, although Texas Instruments TMS320C2x and TMS320C5x processors can address up to 64 Kwords of data memory, instructions using short memory-direct addressing cannot directly specify a complete 16-bit address. Instead, instructions using short memory-direct addressing hold seven bits of address, sufficient for selecting one of 128 (2^7) words. To manage this, the processor divides its data memory into pages of 128 words each. The programmer selects a page by loading a page number into the page pointer register. Subsequent instructions can then make accesses within the 128-word page without requiring a second word of program memory to specify the address. Of course, this technique is efficient only to the extent that the programmer can arrange the program to make a significant number of accesses within one page before needing to access another page.

Chapter 7

Instruction Set

A processor's instruction set is a key factor in determining not only what operations are possible on that processor, but also what operations are natural and efficient. Instructions control how data are sequenced through the processor's data path, how values are read and written to memory, and so forth. As a result, a processor's instruction set can have a profound influence on a processor's suitability for different tasks.

In this chapter we investigate a number of aspects of instruction sets. We look at the types of instructions (and, closely related, the types of registers) commonly found on DSPs and discuss instruction set support for movement of data in parallel with arithmetic and multiply operations. We also explore orthogonality and ease of programming and conclude by examining conditional execution and special function instructions.

7.1 Instruction Types

The paragraphs below discuss the types of instructions (arithmetic, logic, branching, etc.) found on programmable DSP chips. It is important to keep in mind that a processor with more instructions is not necessarily more capable. Indeed, specialized instructions can be a drawback, because they cost silicon area and may not be applicable to the application at hand. This section should provide assistance in understanding the sorts of instructions that are useful in particular applications.

Arithmetic and Multiplication

As discussed previously, arithmetic operations, multiplication, and multiplication-accumulation in particular are critical functions for DSP applications. Most DSP processors provide a rich set of instructions for arithmetic (add, subtract, increment, decrement, negate, round, absolute value) and multiplication. With the exception of the Texas Instruments TMS320C1x processor, all modern DSPs provide multiply-accumulate instructions as well. Typically, these instructions execute in a single instruction cycle.

Some processors (e.g., DSP Group's PineDSPCore and OakDSPCore and Texas Instruments' TMS320C5x) provide instructions to square an operand. These processors' multiply instructions typically use indirectly addressed memory as operands. Other processors that use registers for their multiply operands may not require a special instruction to implement squaring. For example, the Motorola DSP5600x implements squaring with an "MPY X0,X0,A" instruction.

A number of processors provide instructions to support extended-precision arithmetic. For example, signed/signed, signed/unsigned, and unsigned/unsigned multiplication instructions are available on the SGS-Thomson D950-CORE, the Texas Instruments TMS320C54x, and the Zoran ZR3800x. Many processors also support add-with-carry and subtract-with-borrow instructions, both particularly important for extended-precision arithmetic.

Logic Operations

Most DSP processors provide instructions for logic operations, including logical *and*, *or*, *exclusive-or*, and *not*. These find use in error correction coding and decision-processing applications that DSPs are increasingly being called upon to perform. Note that processors may also have bit (or bit-field) manipulation instructions, discussed below.

Shifting

Shifting operations can be divided into two categories: arithmetic and logical. A logical left shift by one bit inserts a zero bit in the least significant bit, while a logical right shift by one bit inserts a zero bit in the most significant bit. In contrast, an arithmetic right shift duplicates the sign bit (either a one or zero, depending on whether the number is negative or not) into the most significant bit. Although people use the term "arithmetic left shift," arithmetic and logical left shifts are really identical: they both shift the word left and insert a zero in the least significant bit.

Arithmetic shifting provides a way of scaling data without using the processor's multiplier. Scaling is especially important on fixed-point processors where proper scaling is required to obtain accurate results from mathematical operations.

Virtually all DSPs provide shift instructions of one form or another. Some processors provide the minimum, i.e., instructions to do arithmetic left or right shifting by one bit. Some processors may additionally provide instructions for two- or four-bit shifts. These can be combined with single-bit shifts to synthesize n-bit shifts, although at a cost of several instruction cycles.

Increasingly, many DSP processors feature a barrel shifter and instructions that use the barrel shifter to perform arithmetic or logical left or right shifts by any number of bits. Examples include the AT&T DSP16xx, the Analog Devices ADSP-21xx and ADSP-210xx, the DSP Group OakDSPCore, the Motorola DSP563xx, the SGS-Thomson D950-CORE, and the Texas Instruments TMS320C5x and TMS320C54x.

Rotation

Rotation can be thought of as circular shifting: shifting where the bits that are shifted off one end of the word "loop around" and are shifted in on the other end. For example, in a left rotate by one bit, the most significant bit is rotated into the least significant bit, and all other bits shift left one position. Rotation finds use in a number of areas, including error correction coding (e.g., for bit interleaving). It can also be used to (slowly) generate bit-reversed addresses on processors that do not have bit-reversal built in to their address generation units.

Most processors provide instructions to rotate a word left or right by one bit. Exceptions include the AT&T DSP16xx, the NEC μPD7701x, the Zilog Z893xx, and the Zoran ZR3800x.

Comparison

Most processors provide a set of *status bits* that provide information about the results of arithmetic operations. For example, status bits commonly include a *zero* bit (set if the result of the last arithmetic operation resulted in a zero value), a *minus* bit (set if the result of the last operation was negative), an *overflow* bit, and so on. Status bits are set as a result of an arithmetic operation and can then be used in conditional branches or conditional execution instructions (discussed below).

In decision-intensive code, a processor may need to rapidly compare a series of values to a known value stored in a register. One way to do this is to subtract the value to be compared from the reference value to set the status bits accordingly. On some processors, this changes the reference value; on others, the result of the subtraction can be placed in another register. On processors with few registers (including most fixed-point DSPs), neither of these is an attractive solution. As an alternative, some processors provide *compare* instructions that effectively perform this subtraction without modifying the reference value or using another register to store the result. An interesting enhancement to the normal compare instruction is the *compare absolute value* instruction, which enables quick determination of whether a number's magnitude falls within a specified range.

Looping

The majority of DSP applications require repeated execution of a small number of arithmetic or multiplication instructions. Because the number of instructions in the inner loop is usually small, the overhead imposed by instructions used to decrement and test a counter and branch to the start of the loop may be relatively large. As a result, virtually all DSP processors provide *hardware looping* instructions. These instructions allow a single instruction or block of instructions to be repeated a number of times without the overhead that would normally come from the decrement-test-branch sequence at the end of the loop.

Hardware looping features are discussed in detail in Chapter 8.

Branching, Subroutine Calls, and Returns

All processors support some form of branching instructions, and most provide subroutine call and return instructions as well. Branch instructions may be called *jump* or *goto* instructions on some processors; similarly, subroutine call instructions may be called jump-to-subroutine instructions. Variations supported by different processors include:

- **Conditional/unconditional**. An unconditional branch always jumps to the specified address, while a conditional branch branches only if the specified condition (or conditions, on some processors) is met.
- **Delayed/multicycle**. A multicycle branch requires several instruction cycles to execute; during this time, the processor cannot do other work. A delayed branch allows the proces-

sor to execute a number of instructions located immediately following the branch before starting execution at the branch destination address. This reduces the number of instruction cycles used for the branch itself. Delayed branches are discussed in detail in Chapter 9.

- **Delayed branch with nullify.** Texas Instruments' TMS320C4x family offers a conditional delayed branch wherein the instructions in the delay slot can be conditionally executed based on whether the branch is taken or not.
- **PC-relative.** A *PC-relative* branch jumps to a location determined by an offset from the current instruction location. This is important in applications that require position-independent code, that is, programs that can run when loaded at any memory address. PC-relative branches are discussed in more detail in Chapter 8.

Conditional Instruction Execution

Conditional instruction execution allows the programmer to specify that an instruction is to be executed only if a specified condition is true. This can be very useful in efficiently implementing decision-intensive code, because it frees the programmer from using branch instructions simply to execute a single instruction as part of an *if-then-else* construct. Conditional execution is especially useful on processors with deep pipelines due to the overhead incurred by branch instructions on such processors.

Some processors build conditional execution into some (or most) of their instructions. For example, most instructions on the Analog Devices ADSP-21xx and ADSP-210xx families allow the programmer to optionally specify the conditions under which the instruction will execute. Other processors, such as the AT&T DSP16xx and the DSP Group PineDSPCore, provide conditional execution of far fewer instructions. Typically, these are instructions that modify the accumulator, e.g., increment, decrement, shift, round, etc. Because the condition codes are built into the instruction opcode, there is no penalty in terms of execution time for using conditional execution.

In contrast, the Texas Instruments TMS320C5x and TMS320C54x families provide a special conditional execution instruction, XC. If the condition specified as XC's argument is true, the next one or two single-word instructions (or the next two-word instruction) are executed. If the condition is false, NOPs are executed instead. The XC instruction itself takes one instruction cycle to execute, but provides a very general conditional execution mechanism.

Special Function Instructions

There are a variety of specialized instructions that are provided on some DSPs. The following are some of the more common examples.

Block floating-point instructions

As discussed in Chapter 3, block floating-point is a form of floating-point arithmetic that is sometimes used on fixed-point processors. Example applications that use block floating-point include speech coding and some implementations of the fast Fourier transform.

Block floating-point makes heavy use of two operations: *exponent detection* and *normalization.* Exponent detection determines the number of redundant sign bits (i.e., the number of leading ones or zeros) in a data value. Normalization refers to exponent detection combined with a left shift that scales the data value so that it contains no redundant sign bits. Exponent detection is sometimes used separately from normalization to determine the maximum exponent in a block of numbers. Once the maximum exponent has been found, a shift of that many bits can be performed on each number in the block to scale them all to the same exponent. Please refer to Chapter 3 for details on block floating-point arithmetic.

DSP processor instructions that support exponent detection and normalization fall into three basic categories:

- **Block exponent detection.** Block exponent detection is used (typically within a hardware loop) to determine the maximum exponent in a block of data. This is a very useful instruction for manipulating arrays of block floating-point data. The Analog Devices ADSP-21xx and Zoran ZR3800x are the only DSP processors that feature block exponent detect instructions.
- **Exponent detection.** Many processors provide instructions that can compute the exponent of a single data value in one instruction cycle. Processors with such instructions include the Analog Devices ADSP-21xx, the DSP Group OakDSPCore, the NEC µPD7701x, the SGS-Thomson D950-CORE, the Texas Instruments TMS320C54x, and the Zoran ZR3800x.
- **Normalization.** Many processors also support normalization (that is, combined exponent detection and shifting), but this support varies widely between processors. AT&T DSP16xx family members (other than the DSP1602 and DSP1605) are the only DSP processors that provide a single-cycle normalize instruction. Most of the processors with single-cycle exponent detection listed in the preceding paragraph can follow the exponent detect instruction with a shift instruction to perform a two-cycle normalization. Finally, a number of other processors, such as the DSP Group PineDSPCore, the Motorola DSP5600x and DSP561xx, and the Texas Instruments TMS320C2x and TMS320C5x, provide *iterative normalization* instructions. These instructions normalize a data value one bit at a time, meaning that normalization of an n-bit number requires n instruction cycles. These instructions usually also compute the exponent of the data value and store this in a processor register.

Bit manipulation instructions

Bit manipulation instructions are useful for both decision processing and error correction coding. They can be broken down into two categories: single-bit manipulation instructions and multibit (or *bit-field*) manipulation instructions.

Single-bit manipulation instructions include bit set, bit clear, bit toggle, and bit test. The Motorola DSP5600x family was one of the first to feature instructions for these operations, although the instructions execute in two instruction cycles. Motorola processors also offer *branch-if-bit-set* and *branch-if-bit-clear* instructions that are useful in control applications.

Bit-field instructions operate on several bits at once. For example, Texas Instruments' TMS320C5x features instructions that use its parallel logic unit to perform a logical operation (*and*, *or*, *not*, *exclusive-or*) on a specified memory location. Similarly, the Motorola DSP561xx family features bit-field test, set, and clear instructions.

AT&T's DSP16xx processors have a special bit manipulation unit that provides the processor with bit-field *extract* and *replace* (*insert* in AT&T parlance) instructions. The former extracts a number of contiguous bits at a specified location within a word and right-justifies them in a destination register. Similarly, the latter takes a set of right-justified bits and uses them to replace the bits at a given position in the destination word. This can be very useful for "protocol processing" in communications applications, for example.

Other special function instructions

There are a wide variety of other special function instructions, some of which are summarized here.

A number of processors provide an *iterative division* instruction that can be used to perform division one bit at a time.

The Analog Devices ADSP-210xx and Texas Instruments TMS320C4x provide a square root seed instruction, which can be used as a basis for finding the square root of a number.

Many processors provide specialized stack operations, push and pop being the most common.

Interrupt enable and disable instructions are used by some processors for control over interrupts.

7.2 Registers

The main registers in a processor are closely coupled to its instruction set, since instructions typically specify registers as source or destination operands, or use them to generate addresses for their source and destination operands. In the paragraphs below, we briefly review the types and functions of registers found in programmable DSPs.

As with many processor features, an abundance of registers usually makes the processor easier to program, but also increases the instruction width and die size of the processor, resulting in a more expensive chip.

Accumulators

Every DSP processor on the market today has at least one accumulator register, although it may be called by a different name on some processors. An accumulator is a register that is at least wide enough to hold the largest ALU or multiplier result produced by the data path and that can be used as a source or destination for arithmetic operations. Some processors, such as the Texas Instruments TMS320C1x, TMS320C2x, and TMS320C5x families, are based on a single primary accumulator. Others, like the AT&T DSP16xx, DSP Group PineDSPCore and OakDSPCore, Motorola DSP5600x and DSP561xx, and Texas Instruments TMS320C54x provide two or more accumulators, which simplifies coding of algorithms that use complex (as opposed to real) numbers. Some floating-point processors provide a large number of extended-precision registers, some of which can be used as accumulators.

General- and Special-Purpose Registers

A number of processors, including the Analog Devices ADSP-210xx, the AT&T DSP32xx, the NEC μPD7701x, the Texas Instruments TMS320C3x and TMS320C4x, and the Zoran ZR3800x, provide a bank of registers (often called a *register file*) for general purpose use. Typically, these registers can be used as operands for most arithmetic and multiplication instructions.

At the opposite extreme, some architectures dedicate registers to certain execution units within the processor. For example, in the Analog Devices ADSP-21xx family the multiplier, ALU, and shifter each has its own dedicated input and output registers. Many DSPs, such as the AT&T DSP16xx, the Zilog Z893xx, and the DSP Group PineDSPCore, have dedicated multiplier output (or product) registers.

On some processors, execution unit input/output registers may also serve as general-purpose registers. For example, the Motorola DSP5600x uses its X0, X1, Y0, and Y1 registers as multiplier inputs, but they can also be used for ALU inputs and other purposes. This increases the processor's flexibility.

Most programmers prefer processors with general-purpose registers that can be used for a variety of tasks over processors that have specialized registers dedicated to particular function units in the data path. However, generality has its price: processors with general-purpose registers tend to be more expensive than those without.

Address Registers

As discussed in Chapter 6 on addressing, address registers (and their associated offset and modifier registers) are used to generate addresses for register-indirect addressing. The number of address registers found on processors ranges from 2 (on the Texas Instruments TMS20C1x) to 22 (on the AT&T DSP32xx).

Many DSP programmers complain that their processors have too few address registers; this is one area where more is usually better.

Other Registers

Other registers found on DSP processors include:

- **Stack pointer**. Typically found on processors that support software stacks, a stack pointer is a register dedicated to indicating the current top of stack location.
- **Program counter**. The program counter holds the address of the next instruction to be fetched.
- **Loop registers**. These hold information regarding hardware loops, such as start and end address and repetition count.

In most cases the programmer's interaction with these other registers is far more limited than with the processor's main registers.

7.3 Parallel Move Support

As discussed in Chapter 5, one of the most important features of a DSP is its ability to perform multiple memory accesses in a single instruction cycle. These memory accesses are often called *parallel moves* because they often occur in parallel with an arithmetic or multiplication operation.

The parallel move formats used by DSPs to achieve this typically take one of two forms, depending on whether or not the data values being moved are directly related to the operands of the ALU or multiply instruction being executed.

The instruction set of DSP Group's PineDSPCore typifies *operand-related* parallel moves. On this processor, only instructions that have multiple source operands can make multiple data memory accesses per instruction cycle, and the accesses are limited to those required by the arithmetic or multiplication operation. For example, a multiply instruction might look like:

```
MPY   (R0),(R4)
```

This instruction fetches the contents of the memory locations pointed to by R0 and R4, multiplies them, and deposits the result in the product register. Note that the two data fetches were made to fetch the operands of the MPY instruction.

In contrast, an example of an *operand-unrelated* parallel move can be found on the Motorola DSP5600x family. On this processor, multiply and ALU operations take their operands from registers that were previously loaded with input values and produce results in the accumulators. In parallel with a multiply or ALU operation, the DSP5600x can access two memory locations (read or write) that may be *unrelated* to the operands of the ALU instruction. A multiply on the DSP5600x might look like:

```
MPY  X0,Y0,A    X:(R0)+,X0      Y1,Y:(R4)+
```

This instruction multiplies X0 and Y0 and places the product in the A accumulator. In parallel, it loads the contents of the X memory location pointed to by R0 into X0, and stores Y1 in the Y memory location pointed to by R4.

Operand-unrelated parallel moves allow the programmer to more efficiently use the memory bandwidth provided by the processor. In the operand-related case, if an instruction does not require two data accesses, those two data accesses are lost to the programmer—even if the underlying hardware has bandwidth to execute them. Operand-unrelated parallel moves allow the programmer to use those two accesses to store results from the previous instruction or to load values in preparation for the next instruction. Operand-unrelated moves are particularly useful in algorithms that require frequent, nonsequential memory accesses, such as the fast Fourier transform.

Processors that provide operand-unrelated parallel moves include the AT&T DSP16xx family, all Analog Devices processors, all Motorola DSP processors, the NEC μPD7701x, the SGS-Thomson D950-CORE, and the Zoran ZR3800x. Processors that use operand-related parallel moves include the AT&T DSP32C and DSP32xx, the DSP Group PineDSPCore and OakDSPCore processors, the Texas Instruments TMS320C2x and TMS320C5x, and the Zilog Z893xx family. Some processors, such as the Texas Instruments TMS320C3x, TMS320C4x, and TMS320C54x support both for different kinds of instructions. The Texas Instruments TMS320C1x does not provide parallel moves at all, as it allows only one data memory access per instruction cycle.

Operand-related and operand-unrelated parallel move support are related to time-stationary and data-stationary assembly coding styles, discussed in Chapter 9.

7.4 Orthogonality

Orthogonality refers to the extent to which a processor's instruction set is consistent. In general, the more orthogonal an instruction set is, the easier the processor is to program. This is because there are fewer inconsistencies and special cases that the programmer must remember. Orthogonality is a subjective topic, and it does not lend itself easily to quantification. This should be kept in mind when reviewing a critique of a processor's orthogonality.

The two major areas that most influence the perception of a processor's orthogonality are the consistency and completeness of its instruction set and the degree to which operands and addressing modes are uniformly available with different operations. As an example of the former, most programmers would describe a processor with an *add* instruction but not a *subtract* instruction as nonorthogonal. As an example of the latter, a processor that allows register-indirect addressing with its add instruction but not with its subtract instruction would also be considered nonorthogonal.

Processors with larger instruction word widths tend to be more orthogonal than processors with smaller instruction words. Fundamentally, this is because orthogonality results from independent encodings of operations and operands within an instruction word. The more bits available in the instruction word, the easier it is to design an orthogonal processor.

As an example, consider a hypothetical processor with 32 instructions that support three operands (for example, "ADD R0,R1,R2", which might add R0 and R1 and place the result in R2). Each operand can be one of eight registers (R0-R7). In addition, suppose the processor supports two data moves in parallel with all instructions, and that the processor has eight address registers that can be used to generate register-indirect addresses for these moves. The processor might also support three address register update modes (no update, post-increment by one, and post-decrement by one).

Table 7-1 shows the breakdown of bits in the instruction encoding. An instruction set encoded in this manner would be quite orthogonal, since every aspect of the instruction set is independently encoded. However, it also requires 30-bit instruction words. Although wider instruction words can provide greater orthogonality and allow the programmer to control more operations per instruction, they also increase required bus and memory widths, thus increasing system cost. In addition, few applications will be able to take full advantage of the instruction set's power and flexibility on every instruction. Every instruction executed that does not make use of all the instruction set's power is an argument for a smaller, less orthogonal, harder to program, but more memory-efficient instruction set.

There are several approaches that are widely used for squeezing more functionality into a smaller instruction word width on DSPs with 16-bit-wide instruction words:

- **Reduced number of operations.** Supporting fewer operations frees instruction word bits for other uses. For example, the AT&T DSP16xx does not have a rotation instruction.

TABLE 7-1. The Number of Bits Required for a Fully Independent Instruction Encoding on a Hypothetical Processor

Purpose	Instruction Set Field	Bits
Instruction code	One of 32 instructions	5
Operands	One of 8 primary source operands	3
	One of 8 secondary source operands	3
	One of 8 destination operands	3
First parallel move	One of 8 primary register-indirect address registers	3
	One of 3 register-indirect update modes	2
	One of 8 source or destination registers	3
Second parallel move	One of 8 secondary register-indirect address registers	3
	One of 3 register-indirect update modes	2
	One of 8 source or destination registers	3
TOTAL		30

- **Reduced number of addressing modes.** Processors with wider instruction word lengths typically feature a rich variety of addressing modes that are available to all instructions. Processors using smaller instruction words typically provide only a few addressing modes, and the number of update modes in register-indirect addressing may be limited. Similarly, the processor may also limit the allowable combinations of operations and addressing modes.
- **Restrictions on source/destination operands.** Motorola's DSP561xx family typifies this solution. For example, the second parallel move in an instruction can only use a specific address register (R3) for address generation. Similarly, the choice of operands in a multiply instruction determines how many parallel reads or writes are allowed.
- **Use of mode bits.** Texas Instruments uses this approach with its TMS320C1x and TMS320C2x processors, and especially on the TMS320C5x. These processors use mode bits or data stored in a variety of registers to partly determine what an instruction does. For example, the TMS230C5x does not have separate arithmetic and logical shift operations. Rather, a shift mode bit in a control register determines whether the single shift instruction is arithmetic or logical. Similarly, the accumulator shift instruction takes its shift count from a special register instead of from a shift count encoded in the instruction word.

Most of these options increase programming difficulty, but the narrower instruction word width usually reduces overall processor and system cost.

7.5 Assembly Language Format

There are two main styles of DSP assembly language: the traditional *opcode-operand* variety and a functional *C-like,* or *algebraic,* syntax. The former expresses instructions in terms of an instruction mnemonic (a brief code indicating the instruction, like MPY) and its operands, for example:

```
MPY   X0,Y0
ADD   P,A
MOV   (R0),X0
JMP   LOOP
```

In contrast, C-like assembly language capitalizes on the syntax of the C programming language and its arithmetic shorthand. The above code fragment might be written in a C-like assembly language as:

```
P = X0 * Y0
A = P + A
X0 = *R0
GOTO LOOP
```

Each style has adherents who claim that their approach is the more natural, clear, or productive. One advantage of the C-like syntax is that algorithms are expressed in assembly language in terms close to their essential mathematical form. It is perhaps telling that one frequently sees opcode-operand assembly language with comments that resemble C-like assembly code, but that one never sees the reverse.

The downside of C-like syntax is that the assembly language is not actually C code, no matter how much it may superficially resemble it. As a result, programmers accustomed to C may find programming in such an assembly language frustrating, since the assembly syntax may suggest the availability of operations that would be possible in C but that are not legal under the assembler.

Finally, it should be noted that the assembly language syntax used to program a processor is really unrelated to the processor's instruction set. That is, there is no reason why a single processor could not have two assemblers, one that uses an opcode-operand syntax and the other that uses a C-like syntax. As long as the two formats generate the same binary opcodes, the two syntaxes are equivalent from the processor's point of view.

Chapter 8

Execution Control

Execution control refers to the rules or mechanisms used in a processor for determining the next instruction to execute. In this chapter, we focus on several features of execution control that are important to DSP processors. Among these are hardware looping, interrupt handling, stacks, and relative branch support.

Execution control is closely related to pipelining, which is discussed in detail in Chapter 9.

8.1 Hardware Looping

DSP algorithms frequently involve the repetitive execution of a small number of instructions—so-called *inner loops* or *kernels*. FIR and IIR filters, FFTs, matrix multiplication, and a host of other application kernels are all performed by repeatedly executing the same instruction or sequence of instructions. DSP processors have evolved to include features to efficiently handle this sort of repeated execution. To understand this evolution, we should first take a look at the problems associated with traditional approaches to repeated instruction execution.

First, a natural approach to looping uses a branch instruction to jump back to the start of the loop. As mentioned in Chapter 7 and as discussed in depth in Chapter 9, branch instructions typically require several instruction cycles to execute. Especially for small loops, the execution overhead of the branch instruction may be considerable.

Second, because most loops execute a fixed number of times, the processor must usually use a register to maintain the loop index, that is, the count of the number of times the processor has been through the loop. The processor's data path must then be used to increment or decrement the index and test to see if the loop condition has been met. If not, a conditional branch brings the processor back to the top of the loop. All of these steps add overhead to the loop and use precious registers.

DSP processors have evolved to avoid these problems via *hardware looping,* also known as *zero-overhead looping*. Hardware loops are special hardware control constructs that repeat either a single instruction or a group of instructions some number of times. The key difference between hardware loops and software loops is that hardware loops lose no time incrementing or decrementing counters, checking to see if the loop is finished, or branching back to the top of the loop. This can result in considerable savings. For example, Figures 8-1(a) and 8-1(b) show an FIR filter implemented in assembly language on two DSP processors, one with software looping and one with hardware looping. The software loop takes roughly three times as long to execute, assuming that all instructions execute in one instruction cycle. In fact, branch instructions usually take several cycles to execute, so the hardware looping advantage is usually even larger.

Except for the Clarkspur Design CD24xx and CD245x cores, the IBM MDSP2780, the Texas Instruments TMS320C1x, and the Zilog Z893xx, all DSPs provide some hardware looping support. However, the exact form of this support may vary widely from one DSP to another. The paragraphs below discuss a number of the more important hardware loop features.

Single- and Multi-Instruction Hardware Loops

Most DSP processors provide two types of hardware loops: single-instruction loops and multi-instruction loops. A single-instruction hardware loop repeats a single instruction some number of times, while a multi-instruction hardware loop repeats a group of instructions some number of times. Of processors supporting hardware looping, all but the Texas Instruments TMS320C2x and its clones provide both single-instruction and multi-instruction hardware loops (the TMS320C2x provides only single-instruction loops). Some processors, such as the Analog Devices ADSP-21xx, allow the use of multi-instruction hardware loops to repeat a single instruction.

Because a single-instruction hardware loop executes one instruction repeatedly, the instruction needs to be fetched from program memory only once. On the second and subsequent executions of the instruction within the repeat loop, the program bus can be used for accessing memory for purposes other than fetching an instruction, e.g., for fetching data or coefficient values that are stored in program memory. Several processors (such as the Texas Instruments TMS320C2x and TMS320C5x) use this mechanism to gain an additional memory access per instruction cycle. On these processors, several types of instructions can execute in a single instruction cycle only if they are executed repeatedly from within a single-instruction hardware loop.

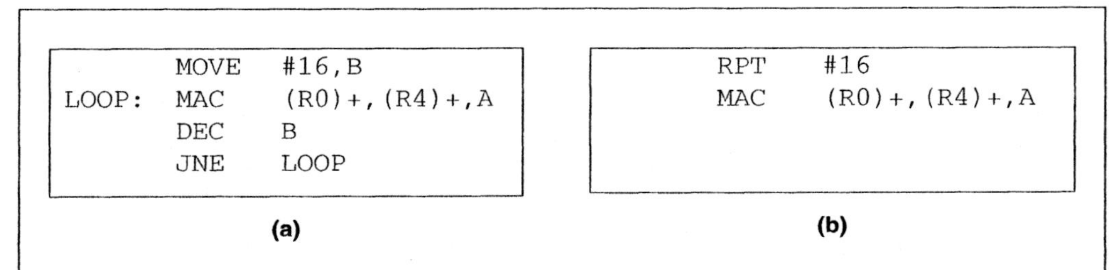

FIGURE 8-1. An FIR filter kernel implemented using a software loop (a) and a hardware loop (b). In the hardware loop, the RPT ("repeat") instruction is executed only once, but automatically repeats the next instruction 16 times. If each instruction takes one instruction cycle (a conservative estimate), the software loop version takes roughly three times longer to execute than the hardware loop version.

In contrast, a multi-instruction hardware loop must refetch the instructions in the block of code being repeated each time the processor proceeds through the loop. Because of this, the processor's program bus is not available to access other data. (One exception to this is the AT&T DSP16xx, which provides a special 15-word buffer that is used to hold instructions being repeated. On the first iteration of the loop, instructions are copied into this buffer. On subsequent passes through the loop the processor fetches instructions from the 15-word buffer, freeing the program bus for data accesses.)

Most processors allow an arbitrary number of instructions to be repeated in a multi-instruction hardware loop. Others, however, only allow a small number of instructions in a multi-instruction loop. For example, the AT&T DSP16xx hardware loop supports a maximum of 15 instructions, and the AT&T DSP32C supports a maximum of 32 instructions. As we noted earlier, most inner loops tend to be small, so such limitations are not unreasonable for many applications.

Loop Repetition Count

Another feature that differentiates processors' hardware looping capabilities is the minimum and maximum number of times a loop can be repeated. Almost all processors support a minimum repetition count of one, and 65,536 is a common upper limit. A few processors, such as the AT&T DSP16xx and DSP Group's PineDSPCore have relatively small upper limits (127 and 256, respectively). Frequently, the maximum number of repetitions is lower if the repetition count is specified using immediate data (as opposed to loading a repetition count register from memory), simply because the processor may place restrictions on the size of immediate data words.

A hardware looping pitfall found on some processors (e.g., Motorola DSP5600x and Zoran ZR3800x) is the following: a loop count of zero causes the processor to repeat the loop the maximum number of times. While this is not a problem for loops whose size is fixed, it can be a problem if the program dynamically determines the repetition count at run time. If zero is a possible repetition count (meaning not to repeat the loop at all), an additional test and branch instruction must be placed before the start of the loop to avoid executing it if the repetition count would have been zero.

Loop Effects on Interrupt Latency

Single-instruction hardware loops typically disable interrupts for the duration of their execution. As a result, system designers making use of both single-instruction hardware loops and interrupts must carefully consider the maximum interrupt lockout time they can accept and code their single-instruction loops accordingly. Alternatives include using multi-instruction loops on single instructions and breaking the single-instruction loop up into a number of smaller loops, one after another, each with smaller repetition counts.

Three DSP processor families from AT&T (the DSP32C, the DSP32xx, and the DSP16xx) disable interrupts even during execution of multi-instruction hardware loops. Programmers of these devices must take extra care to determine the maximum acceptable lockout time in their application and structure their hardware loops accordingly.

Nesting Depth

A nested loop is one loop placed within another. Most applications that benefit from hardware looping need only a single hardware loop. In some applications, it can be both convenient and efficient to nest one hardware loop inside another. For example, the fast Fourier transform requires three nested loops.

The most common approaches to hardware loop nesting are:

- **Directly nestable**. Some processors, including all Motorola DSPs, all Analog Devices DSPs, and the NEC μPD7701x, allow hardware loops to be nested simply by using the hardware loop instruction within the outer loop. Maximum nesting depths range from three (the NEC μPD7701x) to seven (Motorola processors). Processors that feature directly nestable hardware loops usually provide a separate hardware loop stack that holds the loop start and end addresses and repetition count.
- **Partially nestable**. Processors with both single- and multi-instruction hardware loops frequently allow a single-instruction loop to be nested within a multi-instruction loop, even if multi-instruction loops are not nestable themselves. Examples include PineDSPCore from DSP Group and the Texas Instruments TMS320C3x, TMS320C4x, and TMS320C5x processors.
- **Software nestable**. On the Texas Instruments TMS320C3x and TMS320C5x, multi-instruction hardware loops can be nested by saving the state of various looping registers (loop start, loop end, and loop count) and then executing a new loop instruction. Depending on the repetition counts of the inner and outer loops, this may be a better approach than using a branch instruction for the outer loop. (Recall that a branch instruction incurs overhead every iteration through the loop, while saving and restoring loop registers needs to be done only once for each pass through the outer loop.)
- **Nonnestable**. A number of DSPs, including the Texas Instruments TMS320C2x and the AT&T DSP16xx and DSP32xx, do not permit nested hardware loops at all.

Hardware-Assisted Software Loops

An alternative to nesting hardware loops is to nest a single hardware loop within a software loop that uses specialized instructions for software looping. For example, the AT&T DSP16xx and the Texas Instruments TMS320C3x and TMS320C4x support a *decrement-and-branch-if-not-zero* instruction. While this instruction costs one instruction cycle on the TMS320C3x and TMS320C4x and three on the DSP16xx, the number of cycles is less than would be required to execute individual decrement, test, and branch instructions. In many ways, this is a reasonable compromise between the hardware cost required to support nested hardware loops and the execution time cost of implementing loops entirely in software.

8.2 Interrupts

An *interrupt* is an external event that causes the processor to stop executing its current program and branch to a special block of code called an *interrupt service routine*. Typically, this code deals with the source of the interrupt in some way (for example, reading the data that have just arrived on a serial port and storing the in a buffer) and then returns from the interrupt, allowing the processor to resume execution where it left off. All DSP processors support interrupts and most use interrupts as their primary means of communicating with peripherals.

The sections below discuss aspects of interrupt processing and features found on DSPs to support it.

Interrupt Sources

Interrupts can come from a variety of sources, including:

- **On-chip peripherals.** Most processors provide a variety of on-chip peripherals (e.g., serial ports, parallel ports, timers, and so on) that generate interrupts when certain conditions are met.
- **External interrupt lines.** Most processors also provide one or more external interrupt lines that can be asserted by external circuitry to interrupt the processor.
- **Software interrupts.** Also called *exceptions* or *traps*, these are interrupts that are generated either under software control or due to a software-initiated operation. Examples include illegal instruction traps and floating-point exceptions (division by zero, overflow, underflow, and so on).

Both on-chip peripherals and external interrupt lines are discussed in more detail in Chapter 10, "Peripherals."

Interrupt Vectors

As mentioned above, an interrupt causes the processor to begin execution from a certain location in memory. Virtually all processors associate a different memory address with each interrupt. These locations are called *interrupt vectors*, and processors that provide different locations for each interrupt are said to support vectored interrupts. Vectored interrupts simplify programming because the programmer does not have to be concerned with which interrupt source generated the interrupt; this information is implicitly contained in the interrupt vector being executed. Without vectored interrupts, the programmer must check each possible interrupt source to see if it caused the interrupt, increasing interrupt response time.

Typical interrupt vectors are one or two words long and are located in low memory. The interrupt vector usually contains a branch or subroutine call instruction that causes the processor to begin execution of the interrupt service routine located elsewhere in memory. On some processors, interrupt vector locations are spaced apart by several words. This allows brief interrupt service routines to be located directly at the interrupt vector location, eliminating the overhead of branch and return instructions. On other processors, the interrupt vector location does not actually contain an instruction; rather, it contains only the address of the actual interrupt service routine. On these processors, the branch to a different location is mandatory, increasing interrupt service time.

Interrupt Enables

All processors provide mechanisms to globally disable interrupts. On some processors, this may be via special interrupt enable and interrupt disable instructions, while on others it may involve writing to a special control register. Most processors also provide individual interrupt enables for each interrupt source. This allows the processor to selectively decide which interrupts are allowed at any given time.

Interrupt Priorities and Automatically Nestable Interrupts

Most processors provide prioritized interrupts, meaning that some interrupts have a higher priority than others. Our definition of *prioritized* matches that of most processor vendors: if two interrupts occur simultaneously (typically meaning within the same clock cycle), the one with the higher priority will be serviced, and the one with the lower priority must wait for the processor to finish servicing the higher priority interrupt.

By this definition, interrupt priorities are used only to arbitrate *simultaneous* interrupts. They are not used to allow a higher priority interrupt to interrupt a lower priority interrupt that is currently being serviced. When a processor allows a higher-priority interrupt to interrupt a lower priority interrupt that is already executing, we call this *automatically nestable interrupts*. Most DSPs do not support automatically nestable interrupts; Analog Devices and Motorola processors are exceptions. On some processors, an interrupt service routine can explicitly reenable interrupts to allow other interrupts to preempt the current interrupt service routine.

Interrupt Latency

Interrupt latency is the amount of time between an interrupt occurring and the processor doing something in response to it. Interrupt latencies can vary significantly from one processor family to another. Especially for real-time control applications, low interrupt latency may be very important.

Unfortunately, it is often difficult to meaningfully compare interrupt latencies for different processors when working from data books or data sheets. This is because processor vendors often use vague, ill-defined, or contradictory definitions of interrupt latency (or the conditions under which it is measured). Our formal definition of interrupt latency is the minimum time from the assertion of an external interrupt line to the execution of the first word of the interrupt vector that can be guaranteed under certain assumptions. This time is measured in instruction cycles. The details of and assumptions used in this definition are as follows:

- Most processors sample the status of external interrupt lines every instruction cycle. For an interrupt to be recognized as occurring in a given instruction cycle, the interrupt line must be asserted some amount of time *prior* to the start of the instruction cycle; this time is referred to as the *set-up time*. Because the interrupting device has no way of guaranteeing that it will meet the processor's set-up time requirements, we assume that these set-up time requirements are missed. This lengthens interrupt latency by one instruction cycle.

- Once an interrupt line has been sampled by the processor's clock, it is typically passed through several stages of flip-flops to avoid metastability problems. This step is often referred to as *synchronization*. Depending on the processor this can add from one to three instruction cycles to the processor's interrupt latency.
- At this stage the interrupt is typically recognized as valid and pending by the processor. If the processor is not in an interruptible state, interrupt processing is delayed until the processor enters an interruptible state. Examples of noninterruptible states include interrupts being disabled under software control, the processor already servicing another interrupt, the processor executing a single-instruction hardware loop that disables interrupts, the processor executing a multi-instruction-cycle instruction, or the processor being delayed due to wait states when accessing external memory. In our definition of interrupt latency, we assume the processor is in an interruptible state, which typically means that it is executing the shortest interruptible instruction possible.
- The processor must next finish executing instructions that are already being processed in the pipeline. We assume that these instructions are all instructions that execute in a single instruction cycle. In parallel with this, the processor can begin to fetch and process the first word of the interrupt vector. Our definition of interrupt latency includes the time for the first word of the interrupt vector to percolate through to the execute stage of the pipeline.

As described above, some processors do not actually store an instruction word in the interrupt vector location. Rather, on these processors, the interrupt vector location holds only the address of the interrupt service routine, and the processor automatically branches to this address. On such processors, our definition of interrupt latency includes the time to branch to the interrupt service routine. This time is not included for processors that contain instructions at the interrupt vector; this is because the interrupt service routine can reside directly at the interrupt vector in some situations. In other situations, this is not possible, and the time required for the processor to execute a branch instruction must be factored into the interrupt latency calculation as well.

Some processors, notably all Motorola processors and the AT&T DSP32C and DSP32xx, provide *fast* or *quick* interrupts. On Motorola processors, an interrupt is automatically classified as "fast" if the first two instructions at the interrupt vector location are not branch instructions. These two words are inserted directly into the pipeline and executed without a branch, reducing interrupt latency. Fast interrupts are usually used to move data from a peripheral to memory or vice versa.

Interrupt latency is closely tied to the depth and management of the processor's pipeline. Chapter 9 discusses the interactions between interrupts and pipelines in detail.

Mechanisms for Reducing Interrupts

Some processors, such as the Analog Devices ADSP-21xx family, the Texas Instruments TMS320C54x family, and some newer members of the TMS320C5x family, provide "autobuffering" on their serial ports. This feature allows the serial port to save its received data directly to the processor's memory without interrupting the processor. After a certain number of samples have been transferred, the serial port interrupts the processor. This is, in effect, a specialized form of DMA.

8.3 Stacks

Processor stack support is closely tied to execution control. For example, subroutine calls typically place their return address on the stack, while interrupts typically use the stack to save both return address and status information (so that changes made by the interrupt service routine to the processor status word do not affect the interrupted code).

DSP processors typically provide one of three kinds of stack support:

- **Shadow registers**. Shadow registers are dedicated backup registers that hold the contents of key processor registers during interrupt processing. Shadow registers can be thought of as a one-deep stack for the shadowed registers.
- **Hardware stack**. A hardware stack is a special block of on-chip memory, typically only a few words long, that holds selected registers during interrupt processing or subroutine calls.
- **Software stack**. A software stack is a conventional stack using the processor's main memory to store values during interrupt processing or subroutine calls.

Of the three, shadow registers and software stacks are generally the most useful. Shadow registers can be used to provide very low overhead interrupt handling, since they automatically save key processor registers in a single cycle. The key advantage of a software stack over a hardware stack is that its depth can be configured by the programmer simply by reserving an appropriately sized section of memory. In contrast, hardware stacks are usually fairly shallow, and the programmer must carefully guard against stack overflow.

8.4 Relative Branch Support

All DSP processors support branch or jump instructions as one of their most basic forms of execution control.

In some applications (e.g., multitasking PC multimedia systems), a form of branching known as a PC-relative branch can be extremely useful. In PC-relative branching, the address to which the processor is to branch is specified as an offset from the current address. PC-relative addressing is useful for creating *position-independent* programs. A position-independent program is one that can be relocated in memory when it is loaded into the processor's memory. If all of the references to locations in the program code are specified relative to some other location in the program code, then there is no difficulty if the program is loaded into a different part of memory each time it is run. On the other hand, if a program contains absolute references to program memory locations, the program cannot be relocated without modifying these references.

In addition to supporting position-independent code, PC-relative branching can also save program memory in certain situations. In most programs, branches tend to be fairly short. As a result, the offset from the current instruction to the destination instruction frequently does not need to be represented by a full-width instruction word. Instead, it can often be fit into the branch instruction itself, represented as a reduced-width signed offset. For example, DSP Group's OakDSPCore provides a relative branch instruction with a seven-bit signed offset value.

Chapter 9

Pipelining

Pipelining is a technique for increasing the performance of a processor (or other electronic system) by breaking a sequence of operations into smaller pieces and executing these pieces in parallel when possible, thereby decreasing the overall time required to complete the set of operations. Almost all DSP processors on the market today use pipelining to some extent.

Unfortunately, in the process of improving performance, pipelining frequently complicates programming. For example, on some processors pipelining causes certain instruction sequences to execute more slowly in some cases than in others. On other processors, certain instruction sequences must be avoided for correct program operation. Thus, pipelining represents a trade-off between efficiency and ease of use.

This chapter introduces pipeline concepts, provides examples of how they are used in DSP processors, and touches on a number of issues related to pipelining.

9.1 Pipelining and Performance

To illustrate how pipelining increases performance, let's consider a hypothetical processor that uses separate execution units to accomplish the following actions sequentially for a single instruction:

- Fetch an instruction word from memory
- Decode the instruction (i.e., determine what the instruction is supposed to do)
- Read a data operand from or write a data operand to memory
- Execute the ALU or MAC portion of the instruction

Assuming that each of the four stages above takes 20 ns to execute, and that they must be done sequentially, then Figure 9-1 shows the timing of the execution of several instructions. From the figure, we can see that the processor can execute one instruction every 80 ns using this approach. But we can also see that the hardware associated with each stage of instruction execution (instruction fetch, decode, operand fetch, instruction execution) is idle 75 percent of the time (assuming that separate hardware is used for each stage). This is because the processor does not begin processing the next instruction until the current one has completely finished.

A pipelined implementation of this processor starts a new instruction fetch immediately after the previous instruction has been fetched. Similarly, it begins decoding each instruction as soon as the previous instruction is finished decoding. In essence, it overlaps the various stages of execution, as illustrated in Figure 9-2. As a result, the execution stages now work in parallel; while one is fetching the next instruction, another is decoding the preceding instruction, and so on.

From this, we can see that the pipelined approach executes a new instruction every 20 ns, a factor of 4 improvement over the nonpipelined implementation. Figure 9-2 also illustrates a subtle point: an instruction may be spread out over multiple instruction cycles and yet still appear to the programmer to execute in one instruction cycle.

Coincidentally, the hypothetical processor used in this example resembles the Texas Instruments TMS320C3x.

	Clock Cycle							
	1	2	3	4	5	6	7	8
Instruction Fetch	I1				I2			
Decode		I1				I2		
Data Read/Write			I1				I2	
Execute				I1				I2

FIGURE 9-1. Nonpipelined instruction execution. I1 and I2 represent instruction 1 and 2. If each clock cycle takes 20 ns, then one instruction is executed every 80 ns. Note that the hardware associated with each stage in execution (fetch, decode, execute, etc.) sits idle 75 percent of the time.

	Clock (Instruction) Cycle							
	1	2	3	4	5	6	7	8
Instruction Fetch	I1	I2	I3	I4	I5	I6	I7	I8
Decode		I1	I2	I3	I4	I5	I6	I7
Data Read/Write			I1	I2	I3	I4	I5	I6
Execute				I1	I2	I3	I4	I5

FIGURE 9-2. A pipelined implementation of the processor. Instruction 3 is highlighted to show its progress through the pipeline. Note that one instruction is now completed every clock cycle, i.e., every 20 ns, a four-fold improvement over the previous version.

9.2 Pipeline Depth

Although most DSP processors are pipelined, the depth (number of stages) of the pipeline may vary from one processor to another. In general, a deeper pipeline allows the processor to execute faster but makes the processor harder to program. Most processors use three stages (instruction fetch, decode, and execute) or four stages (instruction fetch, decode, operand fetch, and execute). In three-stage pipelines the operand fetch is typically done in the latter part of the decode stage. Analog Devices processors are known for their relatively shallow two-deep pipeline, and the Texas Instruments TMS320C54x processor has a relatively long five-deep pipeline.

9.3 Interlocking

The execution sequence shown in Figure 9-2 is referred to as a *perfect overlap*, because the pipeline phases mesh together perfectly and provide 100 percent utilization of the processor's execution stages. In reality, processors may not perform as well as we have shown in our hypothetical example. The most common reason for this is resource contention. For example, suppose our hypothetical processor takes two instruction cycles (i.e., 40 ns) to write to memory. (This is the case, for example, on AT&T DSP16xx processors.) If instruction I2 attempts to write to memory and instruction I3 needs to read from memory, the second instruction cycle in I2's data write phase conflicts with instruction I3's data read. This is illustrated in Figure 9-3, where the extended write cycle is indicated with a dark border, and the I2 write/I3 read conflict is shaded.

One solution to this problem is *interlocking*. An interlocking pipeline delays the progression of the latter of the conflicting instructions through the pipeline. In the example below, instruction I3 would be held at the decode stage until the read/write stage was finished with instruction I2's write. This is illustrated in Figure 9-4. Of course, holding instruction I3 at decode implies that instruction I4 must be held at the fetch stage. (This may mean that I4 is actually refetched from memory, or it may simply be held in a register.) A side effect of interlocking the

	Instruction Cycle							
	1	2	3	4	5	6	7	8
Instruction Fetch	I1	I2	I3	I4	I5	I6	I7	I8
Decode		I1	I2	I3	I4	I5	I6	I7
Data Read/Write			I1	I2	I2/I3	I4	I5	I6
Execute				I1	I2	I3	I4	I5

FIGURE 9-3. An example of pipeline resource contention. In this example, writes to memory are assumed to take two instruction cycles to complete, meaning that they must be extended by one instruction cycle. The darkened border shows the extended write cycle. Unfortunately, instruction I2's write conflicts with I3's data read (the shaded portion of the figure).

	Instruction Cycle							
	1	2	3	4	5	6	7	8
Instruction Fetch	I1	I2	I3	I4	I4	I5	I6	I7
Decode		I1	I2	I3	I3	I4	I5	I6
Data Read/Write			I1	I2	I2	I3	I4	I5
Execute				I1	I2	NOP	I3	I4

FIGURE 9-4. Use of interlocking to resolve the resource conflict in the previous example. As before, instruction I2 requires two cycles for its read/write stage. The pipeline sequencer holds instruction I3 at the decode stage and I4 at the fetch stage, allowing I2's write to complete. This forces the execute stage to execute an NOP at instruction cycle 6.

pipeline in this way is that there is no instruction to execute at instruction cycle 6, and an NOP is executed instead. This results in a one-instruction cycle penalty when an interlock occurs.

A key observation from this is that an instruction's execution time on a processor with an interlocking pipeline may vary depending on the instructions that surround it. For example, if instruction I3 in our example did not need to read from data memory, then there would be no conflict, and an interlock would not be necessary. This complicates code optimization on an interlocking processor, because it may not be easy to spot interlocks by reading the program code.

It is not hard to envision other resource conflict scenarios. For example, some processors support instructions with long immediate data, that is, data that are part of the instruction but that do not fit in the instruction word itself. Such instructions require an additional program memory fetch to get the immediate data. On many such processors, only one program memory access per instruction cycle is possible. Thus, the long immediate data fetch conflicts with the fetch of the next instruction, resulting in an interlock.

The above examples illustrate pipeline conflicts where interlocking is the only feasible solution; failing to interlock in these examples would produce erroneous results. In other cases, the lack of interlocking can bring about results that may not be erroneous, but may not be what the programmer intended. We'll illustrate this with an example from the Motorola DSP5600x, a processor that makes little use of interlocking.

Like most DSP processors, the DSP5600x provides address registers that are used for register-indirect addressing. An interesting pipeline effect occurs in the following sequence of instructions:

```
MOVE    #$1234,R0
MOVE    X:(R0),X0
```

The first instruction loads the hexadecimal address 1234 into address register R0. The second instruction moves the contents of the X memory location pointed to by address register R0 into register X0. A reasonable expectation would then be that the above instructions move the value stored at X memory address 1234 into register X0. However, due to pipeline effects, the above instructions do not execute in the way one might expect. Understanding why requires a brief examination of the DSP5600x pipeline.

The Motorola DSP5600x uses a three-stage pipeline, made up of fetch, decode, and execute stages. ALU operations (e.g., add, multiply, etc.), data accesses (data memory reads and writes, as in the above example), and register loads are carried out during the execute stage. However, addresses used in data accesses are formed during the *decode* stage. This creates a *pipeline hazard*—a problem due to pipelining that the programmer must be aware of to obtain correct results. In this example, the first MOVE instruction modifies an address register used in address generation, but the modification is not reflected in the address used during the data access for the second MOVE instruction.

Figure 9-5 illustrates this hazard in detail. Note that this figure is slightly different from previous figures as it shows how processor resources (including memory and execution units) are used over time. Such a figure is commonly called a *reservation table*.

Let's assume that at the start of the example R0 contains the hexadecimal value 5678. During instruction cycles 1 and 2, the first move instruction is fetched and decoded, and the second

	Instruction Cycle			
	1	2	3	4
Program Memory	Fetch MOVE #$1234,R0	Fetch MOVE X:(R0),X0	I3	I4
Decode Unit		Decode MOVE #$1234,R0	Decode MOVE X:(R0),X0	I3
Address Generation Unit			Generate data address for X:(R0) access (i.e., 5678)	R0 now equals 1234
Execute Unit			Execute MOVE #$1234,R0	Execute MOVE X:(R0),X0
X Data Memory				Read data from address 5678

FIGURE 9-5. A Motorola DSP5600x pipeline hazard, caused by the use of an address register for register-indirect data access immediately after it was loaded by immediate data in the preceding instruction. The hazard comes from the fact that the register is not loaded until the execute stage, but the data address is generated during the decode stage.

move instruction is fetched. During instruction cycle 3, the second move instruction is decoded. At this time, the address generation unit generates an address for use in the data read of the second move instruction, i.e., address 5678. The first move instruction is being executed, but its results *are not yet available* to the address generation unit. As a result, during instruction cycle 4, the processor reads from X memory address 5678, and register R0 now contains 1234. This is probably not what the programmer intended. To avoid this problem, some other instruction (or even an NOP) can be inserted between the two MOVE instructions to give the pipeline time to get the correct value into R0.

The Texas Instruments TMS320C3x, TMS320C4x, and TMS320C5x processors also face this problem, but use interlocking to protect the programmer from the hazard. For example, the TMS320C3x detects writes to any of its address registers and holds the progression through the pipeline of other instructions that use *any* address register until the write has completed. This is illustrated in Figure 9-6. In this example, an LDI (load immediate) instruction loads a value into an address register, like the first move instruction in the DSP5600x example. The second instruction is an MPYF (floating-point multiply) that uses register-indirect addressing to fetch one of its operands. Because the MPYF instruction uses one of the address registers, interlocking holds it at the decode stage until the LDI instruction completes. This results in the processor executing two NOP instructions.

Processors using interlocking in such situations save the programmer from worrying about whether certain instruction sequences will produce correct output. On the other hand, interlocking also allows the programmer to write slower-than-optimal code, perhaps without even realizing it. This is a fundamental trade-off made by heavily interlocked processors.

9.4 Branching Effects

Another pipeline effect occurs on branches or other changes in program flow. By the time a branch instruction reaches the decode stage in the pipeline and the processor realizes that it must

	Instruction Cycle							
	1	2	3	4	5	6	7	8
Instruction Fetch	LDI	MPYF	I3	I3	I3	I4	I5	I6
Decode		LDI	MPYF	MPYF	MPYF	I3	I4	I5
Data Read/Write			LDI	—	—	MPYF	I3	I4
Execute				LDI	NOP	NOP	MPYF	I3

FIGURE 9-6. The Texas Instruments TMS320C3x approach to solving the pipeline hazard in the previous example. The processor detects writes to address registers and delays the execution of subsequent instructions that use address registers until the write completes. The interlock can be seen during instruction cycles 4 and 5.

begin executing at a new address, the next sequential instruction word has already been fetched and is in the pipeline. One possibility is to discard, or *flush*, the unwanted instruction and to cease fetching new instructions until the branch takes effect. This is illustrated in Figure 9-7, where the processor executes a branch (BR) to a new location. The disadvantage of this approach is that the processor must then execute NOPs for the invalidated pipeline slots, causing branch instructions to execute in multiple instruction cycles. Typically, a *multicycle branch* executes in a number of instruction cycles equal to the depth of the pipeline, although some processors use tricks to execute the branch late in the decode phase, saving an instruction cycle.

	Instruction Cycle							
	1	2	3	4	5	6	7	8
Instruction Fetch	BR	I2	—	—	N1	N2	N3	N4
Decode		BR	—	—	—	N1	N2	N3
Data Read/Write			BR	—	—	—	N1	N2
Execute				BR	NOP	NOP	NOP	N1

FIGURE 9-7. The effect on the pipeline of a multicycle branch (BR) instruction. When the processor realizes a branch is pending during the decode stage of instruction cycle 2, it flushes the remainder of the pipeline and stops fetching instructions. The flushed pipeline causes the branch instruction to execute in four instruction cycles. The processor begins fetching instructions (N1-N4) at the branch destination at cycle 5.

An alternative to the multicycle branch is the *delayed branch*, which does not flush the pipeline. Instead, several instructions following the branch are executed normally, as shown in Figure 9-8. A side effect of a delayed branch is that instructions that will be executed *before* the branch instruction must be located in memory *after* the branch instruction itself, e.g.:

```
BRD   NEW_ADDR    ; Branch to new address.
INST 2            ; These three instructions
INST 3            ; are executed before
INST 4            ; the branch occurs.
```

Delayed branches are so named because, to the programmer, the branch appears to be delayed in its effect by several instruction cycles.

Almost all DSP processors use multicycle branches. Many also provide delayed branches, including the Texas Instruments TMS320C3x, TMS320C4x, and TMS320C5x. The Analog Devices ADSP-210xx, the AT&T DSP32C and DSP32xx, and the Zoran ZR3800x provide only delayed branches.

As with interlocking, multicycle and delayed branches represent trade-offs between ease of programming and efficiency. In the worst case, a programmer can always place NOP instruc-

	Instruction Cycle							
	1	2	3	4	5	6	7	8
Instruction Fetch	BR	I2	I3	I4	N1	N2	N3	N4
Decode		BR	I2	I3	I4	N1	N2	N3
Data Read/Write			BR	I2	I3	I4	N1	N2
Execute				BR	I2	I3	I4	N1

FIGURE 9-8. A delayed branch. Delayed branches do not flush the pipeline. Instead, the instructions immediately following the branch are executed normally. This increases efficiency, but results in code that is confusing upon casual inspection.

tions after a delayed branch and achieve the same effect as a multicycle branch, but this requires more attention on the programmer's part.

Branching effects occur whenever there is a change in program flow, and not just for branch instructions. For example, subroutine call instructions, subroutine return instructions, and return from interrupt instructions are all candidates for the pipeline effects described above. Processors offering delayed branches frequently also offer delayed returns.

9.5 Interrupt Effects

Interrupts typically involve a change in a program's flow of control to branch to the interrupt service routine. The pipeline often increases the processor's interrupt response time, much as it slows down branch execution.

When an interrupt occurs, almost all processors allow instructions at the decode stage or further in the pipeline to finish executing, because these instructions may be partially executed. What occurs past this point varies from processor to processor. We discuss several examples below.

Figure 9-9 illustrates the instruction pipeline during an interrupt on the Texas Instruments TMS320C5x family. One cycle after the interrupt is recognized the processor inserts an INTR instruction into the pipeline. INTR is a specialized branch instruction that causes the processor to begin execution at the appropriate interrupt vector. Because INTR behaves like a multi-instruction branch, it causes a four-instruction delay before the first word of the interrupt vector (V1 in the diagram) is executed. Note that the instructions that were at or past the decode stage in the pipeline when the interrupt was recognized (I3-I5) are allowed to finish their execution. Instruction I6 is discarded, but will be refetched on return from interrupt. Note also that the figure does not reflect time for interrupt synchronization, as discussed in Chapter 8.

Figures 9-10 and 9-11 show two of several possible interrupt processing scenarios on the Motorola DSP5600x. A number of differences from the TMS320C5x are immediately apparent.

	Instruction Cycle									
	1	2	3	4	5	6	7	8	9	10
Instruction Fetch	I4	I5	I6	—	—	—	V1	V2	V3	V4
Decode	I3	I4	I5	INTR	—	—	—	V1	V2	V3
Data Read/Write	I2	I3	I4	I5	INTR	—	—	—	V1	V2
Execute	I1	I2	I3	I4	I5	INTR	NOP	NOP	NOP	V1

Interrupt processing begins

FIGURE 9-9. TMS320C5x instruction pipeline while handling an interrupt. The processor inserts the INTR instruction into the pipeline after the instructions that were in the pipeline. INTR is a specialized branch that flushes the pipeline and jumps to the appropriate interrupt vector location. The first word of the interrupt vector is fetched in cycle 7 and executed in cycle 10.

First, the processor takes two instruction cycles after recognizing the interrupt to begin interrupt processing. Second, the DSP5600x does not use an INTR instruction to cause the processor to execute at the vector address. Instead, the processor simply begins fetching from the vector location after the interrupt is recognized. However, at most two words are fetched starting at this address. If one of the two words is a subroutine call (Figure 9-10), the processor flushes the previously fetched instruction and then branches to the long interrupt vector. If neither word is a subroutine call, the processor executes the two words and continues executing from the original program, as shown in Figure 9-11. This second form of interrupt routine is called a *fast interrupt.* Interrupts are discussed in more detail in Chapter 8.

	Instruction Cycle						
	1	2	3	4	5	6	7
Instruction Fetch	I3	I4	JSR	—	V3	V4	V5
Decode	I2	I3	I4	JSR	—	V3	V4
Execute	I1	I2	I3	I4	JSR	—	V3

Interrupt processing begins

FIGURE 9-10. Motorola DSP5600x interrupt processing. The first word of the interrupt vector (V1) is a subroutine call (JSR) instruction, and the second word (V2) is not fetched. This causes a flush of the pipeline, resulting in the loss of one instruction cycle. The interrupt service routine at the destination JSR address is fetched in cycle 5 and executes in cycle 7.

	Instruction Cycle						
	1	2	3	4	5	6	7
Instruction Fetch	I3	I4	V1	V2	I5	I6	I7
Decode	I2	I3	I4	V1	V2	I5	I6
Execute	I1	I2	I3	I4	V1	V2	I5

↑ **Interrupt processing begins**

FIGURE 9-11. "Fast" interrupt processing on the Motorola DSP5600x. In this example, the two words in the interrupt vector (instructions V1 and V2) are not subroutine calls, so the two words are simply inserted into the pipeline with no overhead.

9.6 Pipeline Programming Models

The examples in the sections above have concentrated on the instruction pipeline and its behavior and interaction with other parts of the processor under various circumstances. In this section, we briefly discuss two major assembly code formats for pipelined processors: time-stationary and data-stationary.

In the time-stationary programming model, the processor's instructions specify the actions to be performed by the execution units (multiplier, accumulator, and so on) during a single instruction cycle. A good example is the AT&T DSP16xx family where a multiply-accumulate instruction looks like:

```
a0=a0+p    p=x*y    y=*r0++    p=*pt++
```

The accumulator adds the previous product (register P) to its current contents, the multiplier multiplies the X and Y registers and places the result in the product register, and the X and Y registers are loaded with the values at the memory locations pointed to by registers R0 and PT. Note that each portion of the instruction operates on separate operands.

Data-stationary programming specifies the operations that are to be performed, but not the exact times during which the actions are to be executed. As an example, consider the following AT&T DSP32xx instruction:

```
a1 = a1 + (*r5++ = *r4++) * *r3++
```

The values in the memory locations pointed to by registers R3 and R4 are fetched and multiplied. The value that was pointed to by register R4 is written back to the memory location pointed to by register R5. The result of the multiplication is accumulated in register A1. Note that this style of code tracks a single set of operands through a sequence of operations.

Unlike the DSP16xx instruction, the data-stationary approach uses operands that refer to memory directly. Instead of multiplying two registers, the instruction multiplies the contents of two memory locations. The values of these memory locations are fetched and brought to the multiplier, but the programmer does not specify this operation directly as in the time-stationary case. The processor schedules the sequence of memory accesses and carries them out without explicit direction.

The data-stationary and time-stationary approaches each have their adherents. In general, the data-stationary model is easier to read but not as flexible as the time-stationary approach.

Chapter 10

Peripherals

Most DSP processors provide a good selection of on-chip peripherals and peripheral interfaces. This allows the DSP to be used in an embedded system with a minimum amount of external hardware to support its operation and interface it to the outside world. In this chapter we discuss peripherals and peripheral interfaces commonly found on DSP processors.

10.1 Serial Ports

A serial interface transmits and receives data one bit at a time. In contrast, parallel ports send and receive data in parallel, typically 8, 16, or 32 bits at a time. Although serial interfaces are not as fast as parallel interfaces (in terms of the number of bits transferred per second), they require far fewer interface pins (as few as three or four for a complete serial interface).

Serial ports on DSP chips are used for a variety of applications, including:

- Sending and receiving data samples to and from A/D and D/A converters and codecs
- Sending and receiving data to and from other microprocessors or DSPs
- Communicating with other external peripherals or hardware

There are two main categories of serial interfaces: synchronous and asynchronous. A *synchronous* serial port transmits a bit clock signal in addition to the serial data bits. The receiver uses this clock to decide when to sample the received serial data.

In contrast, *asynchronous* serial interfaces do not transmit a separate clock signal. Rather, they rely on the receiver deducing a clock signal from the serial data itself, which complicates the serial interface and limits the maximum speed at which bits can be transmitted. Asynchronous serial ports are typically used for RS-232 or RS-422 communications. Although there are DSPs that provide asynchronous serial ports (for example, the Motorola DSP5600x and some members of the NEC μPD7701x family), they are rare and are not discussed further here.

Almost all DSP processors provide one or more on-chip synchronous serial interface ports. Typically, each port allows transmission and reception of information over a serial channel. The paragraphs below describe aspects of synchronous serial interfaces.

Data and Clock

As mentioned above, a synchronous serial transmitter sends a *bit clock* signal that the receiving serial interface uses to determine when data bits are valid. Figure 10-1 shows an example data waveform and bit clock. All synchronous serial interfaces assume that the transmitter changes the data on one clock edge (either the rising or falling edge), and that the data are stable (not changing) on the other clock edge. In the figure, data change on the rising edge of the clock and are read by the receiver on the falling edge. However, different serial interfaces or peripherals may assume different clock/data relationships. As a result, some DSP chip serial ports allow the programmer to choose the *clock polarity*, i.e., which edge controls when data change.

Most, but not all, serial interface circuits assume that a positive voltage (typically near 3.3 or 5.0 V) on the serial data line represents a logic "1," and that a low voltage (near 0 V) represents a logic "0." Serial ports on some DSP chips allow the programmer to select data polarity as well.

The order in which bits are transmitted is another issue. In Figure 10-1, we have assumed that words are eight bits long and are transmitted with the least significant bit (LSB) first. Not all devices that use serial protocols to communicate accept this convention; some send the most significant bit (MSB) first. As a result, some DSP serial interfaces allow selection of shift direction, i.e., LSB or MSB first.

Serial ports support a variety of word lengths, i.e., the number of bits in a transmitted data word. Common values are 8 bits (typically used in telephony applications) and 16 bits (used in digital audio), but some serial ports support other word lengths as well.

Frame Synchronization

Synchronous serial interfaces use a third signal in addition to data and clock called the *frame synchronization*, or *frame sync*, signal. This signal indicates to the receiver the position of the first bit of a data word on the serial data line. Frame sync is known by a variety of names; some manufacturers call it a "word sync" or "start of word" signal.

Two common formats for frame sync signals are *bit length* and *word length*. These names refer to the duration of the frame sync relative to the duration of one data bit. A bit-length frame

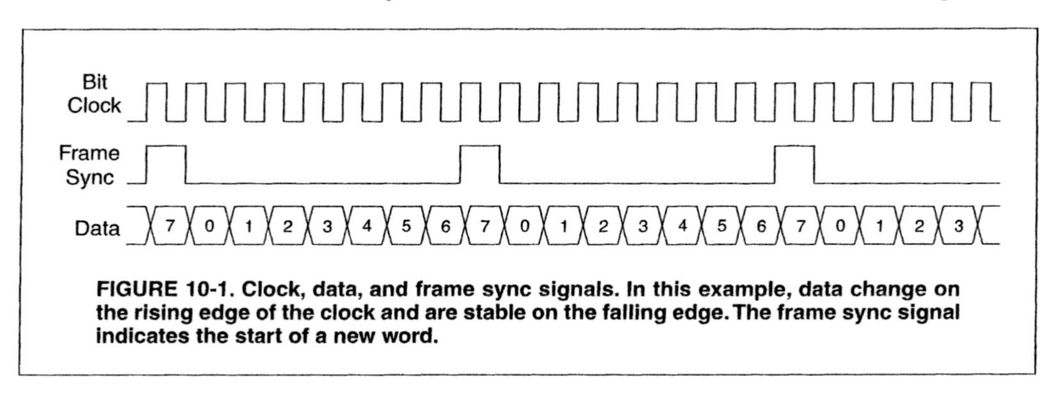

FIGURE 10-1. Clock, data, and frame sync signals. In this example, data change on the rising edge of the clock and are stable on the falling edge. The frame sync signal indicates the start of a new word.

sync signal lasts one bit time and typically occurs one bit before the first transmitted bit of the data word, although it can also occur simultaneously with the first transmitted bit of the data word. The frame sync signal shown in Figure 10-1 is a bit-length frame sync. In contrast, a word-length frame sync lasts the entire length of the data word. An example word-length frame sync is shown in Figure 10-2. Most DSPs support bit-length frame sync signals, and a good number also support word-length frame syncs.

An added complication is that some external devices may expect an inverted frame sync, meaning a frame sync signal that is normally high but pulses low at the start of a data word. Relatively few DSPs allow choice of frame sync polarity.

Multiple Words Per Frame

Some serial ports can handle only one word per frame sync pulse. A little thought reveals that a frame sync signal may not be needed for every word. (In fact, if transfers are periodic, the frame sync is actually needed only once at initialization time.) As a result, some DSP serial ports support multiple words per frame, as shown in Figure 10-3. This is particularly useful for stereo A/D and D/A converters, since they use two words per frame: a left channel sample and a right channel sample. The use of a word-length frame sync signal is natural in this case, since the state of the frame sync (high or low) can be used to determine the source or destination channel (left or right).

Independent Receive and Transmit Sections

A serial port on a DSP is usually made up of a receive section and a transmit section. On some DSPs, the receive and transmit sections are independent (i.e., the transmitter and receiver

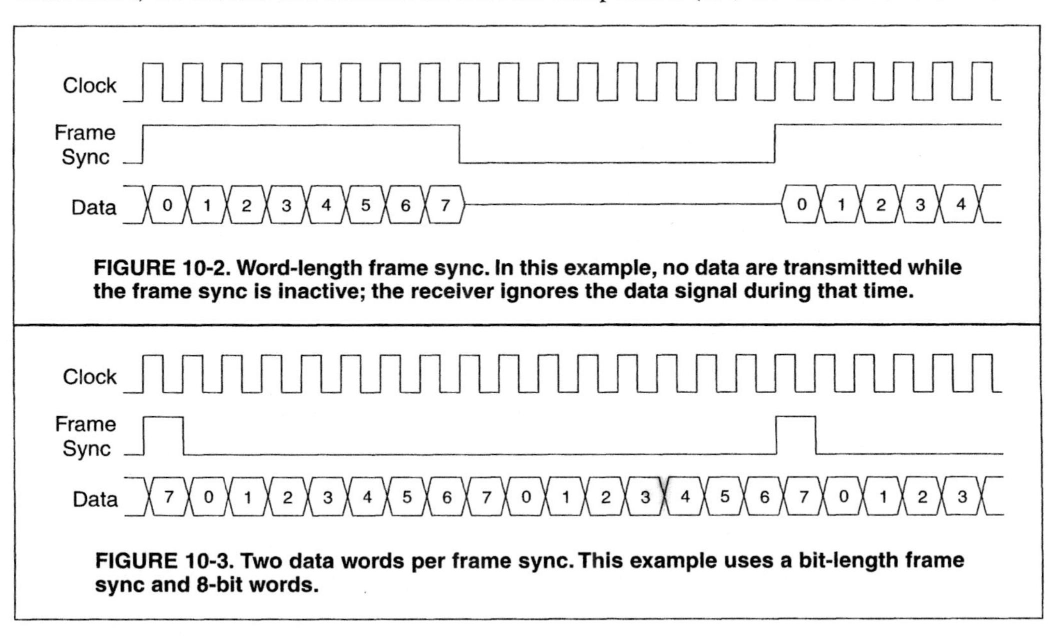

FIGURE 10-2. Word-length frame sync. In this example, no data are transmitted while the frame sync is inactive; the receiver ignores the data signal during that time.

FIGURE 10-3. Two data words per frame sync. This example uses a bit-length frame sync and 8-bit words.

may have independent clock and frame sync pins), while on others, the two may be ganged together and use separate data pins but common clock and frame sync lines. The former arrangement provides much greater flexibility, but this flexibility is not needed in all situations.

Serial Clock Generation

When any two devices are connected via a synchronous serial link, both devices must agree on where the clock will come from. One of the two devices must generate the clock, or the clock must be generated by an external, third device.

Some DSPs provide circuitry for generating a serial bit clock for use with their synchronous serial port. This is usually called *serial clock generation* or *baud rate generation* support. Serial clock generation support varies widely among DSPs; some may not be able to generate their own clocks at all, others may be able to generate a transmit clock but not a receive clock, and still others may have multiple clock generators.

Serial clock generators usually produce a serial bit clock by dividing down the DSP's master clock by some value. The degree of control the user has over the generated clock frequency varies widely; some DSPs provide extensive control, while others allow the user to choose only from a limited set of frequencies.

In general, serial clock generators cascade together a *prescaler* and a programmable *counter* to divide down the master clock frequency. The prescaler is also a counter, but usually can only divide the clock down by a few hard-wired divisors from which the programmer must choose. For example, the Motorola DSP5600x family allows the user to select a prescaler of divide-by-1 (no change) or divide-by-8, and the prescaler on AT&T DSP16xx processors supports divider ratios of 4, 12, 16, or 20.

The counter portion of the clock generator is usually a programmable down counter, clocked with the output of the prescaler, that produces a pulse upon reaching zero. At that time, the counter reloads its count value and begins counting down again.

Some DSP serial clock generators do not include a programmable counter, forcing the programmer to select one of a few possible clock rates. Others provide only a counter (but no prescaler), which may not be able to reduce the clock frequency to the desired frequency.

Time Division Multiplex (TDM) Support

Synchronous serial ports are sometimes used to connect more than two DSP processors. This is typically done using *time division multiplexing*, or TDM. In a TDM network, time is divided into time slots. During any given time slot, one processor can transmit, and all other processors must listen.

Figure 10-4(a) shows a typical TDM network. Processors communicate over a three-wire bus: a data line, a clock line, and a frame sync line. Each DSP's frame sync and clock line is tied to the appropriate line on the bus, and the DSP's transmit and receive lines are *both* tied to the data line.

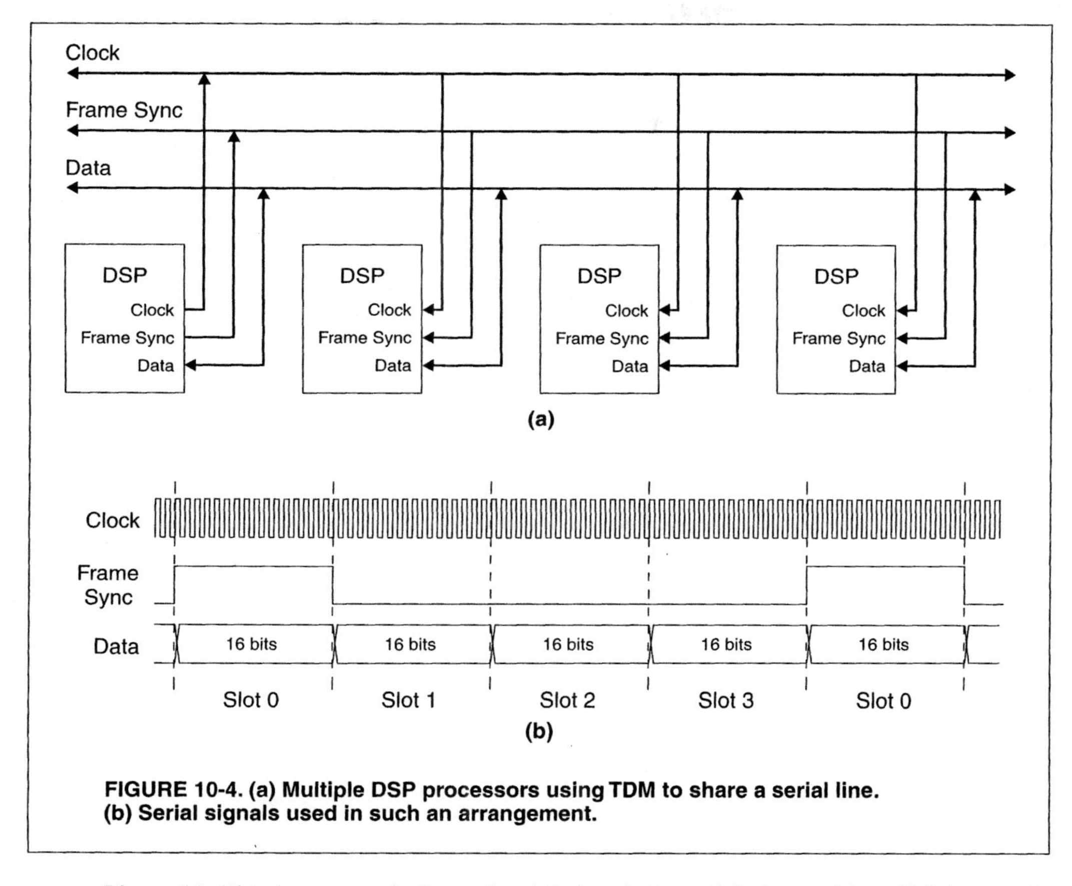

FIGURE 10-4. (a) Multiple DSP processors using TDM to share a serial line. (b) Serial signals used in such an arrangement.

Figure 10-4(b) shows a typical set of serial signals that might be used in a TDM network. One processor (or external circuitry) is responsible for generating the bit clock and frame sync signals. The frame sync signal is used to indicate the start of a new set of time slots. After the frame sync, each processor must keep track of the current time slot number and transmit only during its assigned slot. A transmitted data word (16 bits in the figure) might contain some number of bits to indicate the destination DSP (e.g., two bits for four processors) with the remainder containing data. Another approach uses a secondary data line to transmit the source and destination address in parallel with the data word.

At a minimum, TDM support requires that the processor be able to place its serial port transmit data pin in a high-impedance state when the processor is not transmitting. This allows other DSPs to transmit data during their time slots without interference. Some DSP serial ports have additional support, such as a register that causes the serial port to transmit during a particular time slot. Without this support, the processor must receive all data sent to all other processors and throw away the data not addressed to it in order to know when it is its turn to transmit.

Companding Support

Synchronous serial ports are frequently used to interface with off-chip *codecs*, which are special purpose A/D and D/A converters used in telephony and low-fidelity voiceband applications. Codecs usually use eight-bit data words representing compressed 14- or 16-bit samples of audio data. The compression scheme is typically based on one of two standards, μ-law (in the United States) or A-law (in Europe). The process of compressing and expanding the data is called *companding* and is usually done in software via lookup tables. Some DSPs, notably the Analog Devices ADSP-21xx family and the Motorola DSP56156, provide companding support built into their serial interfaces. This feature can save both execution time and memory by relieving the processor of the table-lookup functions. Interestingly, the AT&T DSP32C and DSP32xx have these format conversions built into their data paths.

Implications for the System Designer

As illustrated above, there are a large number of variations on basic synchronous serial communication. These range from the mundane (e.g., whether the first or last bit in a word is transmitted first, word-length versus bit-length frame sync, clock polarity) to the obscure (e.g., TDM and companding support). The serial ports found on some DSP processors are highly configurable; others are not. This suggests that the system designer should carefully check the compatibility between a processor's supported serial port data formats and the peripherals with which it is intended to be used before making design decisions that may later prove painful if an incompatibility is discovered.

10.2 Timers

Almost all DSPs provide programmable timers. These are often used as a source of periodic interrupts (e.g., as a "clock tick" for a real-time operating system), but other timer applications are possible as well. For example, some DSPs include an output pin that provides a square wave at the frequency generated by the timer. Because this frequency is under software control, this output can be thought of as a software-controlled oscillator, useful in implementing phase-locked loops.

Fundamentally, a timer is much like a serial port bit clock generator: it consists of a clock source, a prescaler, and a counter, as shown in Figure 10-5. The clock source is usually the DSP's master clock, but some DSPs allow an external signal to be used as the clock source as well.

The purpose of the prescaler is to reduce the frequency of the clock source so that the counter can count longer periods of time. It does this by dividing the source clock frequency by one of several selectable values. For example, in the Analog Devices ADSP-21xx family, the prescaler uses an 8-bit value, dividing the input clock by a value from 1 to 256. In the AT&T DSP16xx family, the prescaler uses a 4-bit value N to divide the input clock by $2^{(N+1)}$, i.e., by powers of 2 from 2 through 65,536.

The counter then uses this prescaled signal as a clock source, typically counting down from a preloaded value on every rising clock edge. The counter usually interrupts the processor upon reaching zero. At this point, the counter may reload with the previously loaded value, or it may stop, depending on the configuration of the timer.

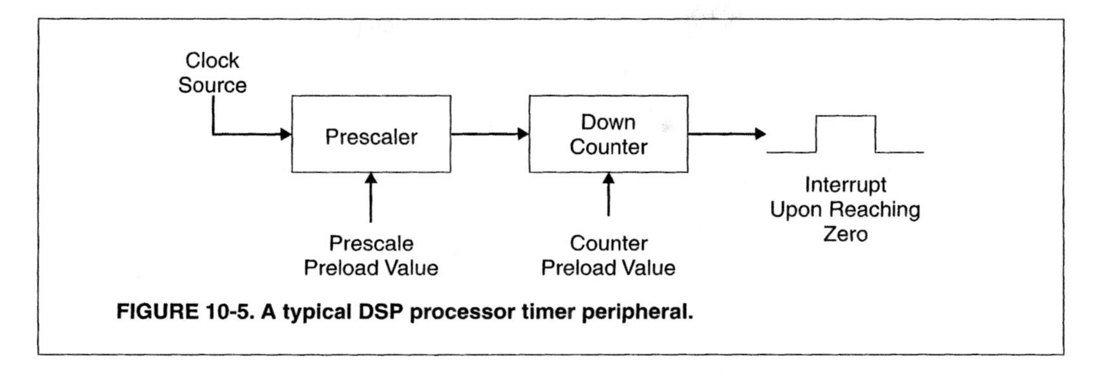

FIGURE 10-5. A typical DSP processor timer peripheral.

Most counter registers on DSP chips are 16 bits wide, which provides a fairly large range of interrupt frequencies that can be generated. On most chips, the user can read the value stored in the counter register. This allows the DSP to read the counter register at the start and end of some event and thus calculate the event's duration.

In the Motorola DSP56000 and DSP56001, the timer is shared with the asynchronous serial clock generator. To use both the timer and the serial clock generator on these DSPs, the user must select a common frequency that is acceptable for both applications.

As mentioned above, some DSPs optionally make the clock signal generated by the timer available on an output pin. This can be useful either for a software-controlled phase-locked loop or simply as a variable frequency synthesizer. Most DSPs, however, have timers that are only capable of generating interrupts and do not have output pins.

10.3 Parallel Ports

A parallel port transmits and receives multiple data bits (typically 8 or 16 bits) at a time. Parallel ports can usually transfer data much faster than serial ports but require more pins to do so. In addition to some number of data lines, a parallel port typically includes *handshake* or *strobe* lines. These lines indicate to an external device that data have been written to the port by the DSP, or vice versa.

Some DSPs use the processor's main data bus as a parallel port. They typically do this by reserving a special section of their address space (or use a special instruction) for I/O accesses over the external bus. When the special memory address is accessed (or the I/O instruction executed), a special strobe or handshake pin is asserted and the processor performs an external read or write bus cycle, as appropriate. An external device monitoring this pin then knows that it should read from or write to the processor's external data pins. This approach is quite common and is used by most Texas Instruments DSPs.

In contrast, other DSPs, such as the AT&T DSP16xx family, provide parallel ports that are separate from the processor's external bus interface. These pins may be dedicated to the parallel port, or they may be shared with other peripherals. Separating the parallel port from the processor's external bus interface can simplify interfacing to external devices.

10.4 Bit I/O Ports

Bit I/O refers to a parallel I/O port wherein individual pins can be made inputs or outputs on a bit-by-bit basis. Unlike traditional parallel I/O ports, bit I/O ports do not usually have associated strobe or handshake lines and may not have any interrupt support. Rather, the processor must poll the bit I/O port to determine if input values have changed. Bit I/O is often used for control purposes but is sometimes used for data transfer, too.

Most DSPs provide at least one or two bit I/O lines, although some provide dozens. On some processors, bit I/O ports are shared with other peripheral pins. For example, on the Motorola DSP5600x, some of the pins associated with a serial port may be configured for use as bit I/O if the application does not require use of the serial port. On other processors, the bit I/O port may use dedicated pins.

Many processors include special conditional instructions whose execution depends on the state of the bit I/O pin. For example, most Texas Instruments fixed-point processors include a bit I/O input pin called "BIO" and also include a BIOZ instruction ("branch if I/O zero") that conditionally branches if the BIO pin is zero.

Both the Texas Instruments TMS320C14 and the AT&T DSP16xx family of DSPs have particularly sophisticated bit I/O ports. In addition to being able to individually configure the lines as inputs or outputs as well as set and clear the bits, these bit I/O units continually check to see if the input bits match a preprogrammed pattern. On the DSP16xx, the result of this comparison sets the state of four flags: all bits match the pattern, some bits match the pattern, some bits do not match the pattern, and no bits match the pattern. These flags can then be used in conditional instructions. For example, using the bit I/O unit, a DSP16xx instruction can conditionally increment a register if one of two specified bits is set. The TMS320C1x bit I/O unit can interrupt the processor when the input pattern matches the stored pattern.

10.5 Host Ports

Some DSPs provide a *host port* for connection to a general-purpose microprocessor or another DSP. Host ports are usually specialized 8- or 16-bit bidirectional parallel ports that can be used for data transfer between the DSP and host processor. In some cases, the host port can also be used to control the DSP (e.g., force it to execute instructions or interrupt service routines, read or write registers and memory, etc.). It may also be used for bootstrap loading.

DSPs providing host ports include all Motorola processors, some Analog Devices processors, the NEC μPD7701x, and recent Texas Instruments fixed-point processors. DSP Group provides a host port macrocell for their PineDSPCore and OakDSPCore.

10.6 Communications Ports

A communications port, or *comm port*, is a special parallel port intended for multiprocessor communication. A comm port differs from a host port or generic parallel port in two ways. First, it is intended for interprocessor communication between the same types of DSPs (as opposed to connecting different types of processors or connecting to peripherals). Second, comm ports generally do not provide special functions for controlling the DSP processor, as are often found on host ports.

Currently the Texas Instruments TMS320C4x and the Analog Devices ADSP-2106x are the only DSPs with comm ports. Both the TMS320C40 and the ADSP-2106x have six comm ports, while the TMS320C44 has four; the TMS320C4x's are each eight bits wide, while the ADSP-2106x's are four bits wide. Since both processors have a native data word length of 32 bits, the comm ports on both processors provide FIFOs for fragmenting and reassembling larger words when they are transmitted across the relatively narrow comm ports.

10.7 On-Chip A/D and D/A Converters

Some DSPs that are targeted at speech applications (for example, digital cellular telephones or tapeless answering machines) have on-chip A/D and D/A converters (also called codecs, although they may not produce the companded samples usually associated with a codec).

There are a number of criteria useful in evaluating on-chip codecs, including:

- **Resolution (in bits) of samples.** Typical codecs provide 16 bits of resolution.
- **Sampling rate.** Most codecs produce and consume samples at 8 Ksamples/s. Some, like those found on the Motorola DSP561xx family, produce or consume samples at rates between 16 and 24 Ksamples/s. In some cases, the DSP processor may have to spend cycles pre- and postprocessing the data to resample them to a desired sample rate.
- **Signal to noise-plus-distortion ratio.** This number is the ratio of signal power (typically a sinusoid) to the power of codec-introduced noise and distortion, and provides an overall measure of codec fidelity. It increases as the input signal increases, but eventually peaks at a maximum and then falls off rapidly as clipping occurs. Vendors usually quote the single best-case number, which is typically in the neighborhood of 65 dB. Values may be specified for both the input and output portions of the codec.
- **Number of analog input channels.** Most on-chip codecs provide an analog multiplexer circuit that allows the DSP to switch between two off-chip analog sources under software control. (Note that this is not the same as having two codecs on-chip!)
- **Programmable output gain.** Codecs typically provide programmable output gain, but the gain range and step size may vary.
- **Analog power-down.** A significant power management feature is the ability to power-down the analog portion of the codec when it is not in use.

Two families of recently designed processors with on-chip A/D and D/A support include the Motorola DSP561xx and the Analog Devices ADSP-21msp5x. Both parts provide codecs with 16-bit resolution at an effective sample rate of 8 kHz, although Motorola's codec requires extra processing on the part of the DSP, which reduces the number of cycles available for the user's application.

10.8 External Interrupts

Most DSPs provide external interrupt lines. These are pins that an external device can assert to interrupt the processor. Interrupt lines fall into two categories:

- **Edge-triggered.** A rising (or falling) edge on the pin asserts the interrupt.
- **Level-triggered.** A high or low level on the pin asserts the interrupt.

Some DSPs provide only level- or edge-triggered interrupt lines, while others allow the interrupt line to be configured to be either. In general, edge-triggered interrupts are easier to work with than level-triggered ones, since level-triggered interrupts may place restrictions on the minimum duration that the interrupt line must be asserted for it to be recognized.

The number of interrupt lines varies between processors, but most provide between one and four external interrupt lines. In general, DSP cores need to (but often do not) provide a greater number of interrupt lines than packaged processors. This is because most packaged DSP processors include a number of peripherals on-chip, each with its own internal interrupt lines. In contrast, most DSP cores do not include on-core peripherals. When on-chip peripherals are added off-core, they often need their own interrupt lines.

10.9 Implications for the System Designer

Peripherals and peripheral interfaces are important considerations when selecting a DSP processor. The power of the peripheral interfaces provided by different DSP processors can have a significant impact on their suitability for a particular application. However, overly-flexible or unneeded I/O interfaces increase processor cost; as with many DSP processor capabilities, more is not necessarily better. On-chip peripherals and peripheral interfaces should be carefully evaluated along with other processor features, such as arithmetic performance, memory bandwidth, and so on.

Chapter 11

On-Chip Debugging Facilities

Debugging is one of the most difficult stages of the design process. This is especially true for real-time embedded systems, where access to the components being debugged may be quite limited, and where it may be difficult to run the system other than in real-time. As a result, DSP processor features that facilitate debugging have become more important as systems have increased in complexity.

One of the most important innovations in this area is scan-based in-circuit emulation (ICE), which combines debugging circuitry on the processor with dedicated test/debug pins to allow debugging of the DSP's operation while it is installed in the target system. Scan-based debugging has become quite popular on DSP processors in recent years, although several other debugging approaches (notably pod-based in-circuit emulation and monitor-based debugging) are still in use.

This chapter focuses on scan-based on-chip debugging facilities. Other debugging techniques, such as pod-based emulation, are discussed in Chapter 16, "Development Tools."

11.1 Scan-Based Debugging/Emulation Facilities

Scan-based emulation uses a small number of dedicated pins (as few as four) and dedicated circuitry on the processor for in-circuit debugging and emulation, as shown in Figure 11-1. Scan-based interfaces typically connect to a host computer (PC or workstation) with a special interface card. The debugging software running on the PC or workstation allows the user to download programs, examine and modify registers and memory, set and clear breakpoints, and perform other functions on the DSP while it is installed in the target system.

Scan-based emulation is much less intrusive than other approaches to debugging or in-circuit emulation. It does not require the processor to be physically removed from the target system and replaced with an emulator processor surrounded by support hardware. Instead, the debugging interface card in the host computer connects to a dedicated connector in the target system, resulting in far fewer physical and electrical problems.

Another advantage of scan-based debugging is that dedicated on-chip debugging circuitry provides visibility into aspects of the processor's execution that are not available with conven-

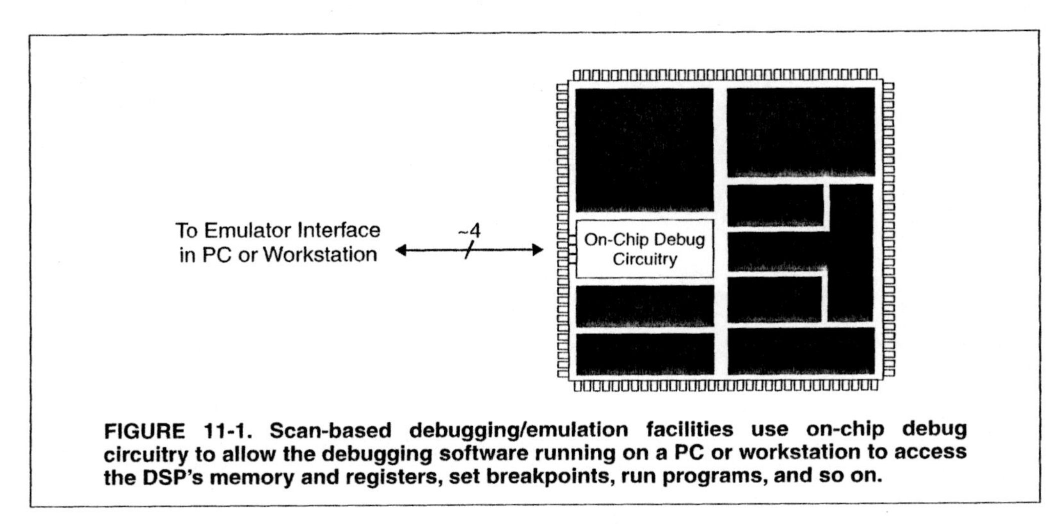

FIGURE 11-1. Scan-based debugging/emulation facilities use on-chip debug circuitry to allow the debugging software running on a PC or workstation to access the DSP's memory and registers, set breakpoints, run programs, and so on.

tional pod-based techniques. For example, Motorola's scan-based debugging circuitry allows the user to examine the contents of the processor's pipeline after reaching a breakpoint.

Scan-based emulation is frequently associated with *boundary-scan*, a testability technique that became popular in the 1980s. Boundary-scan uses a small number of device pins (typically three or four) running a simple serial protocol to allow an external device to observe the values on all of the input pins of an IC and to force values onto the output pins of the IC. This provides a noninvasive means for testing the interconnections between chips: using the serial scan interfaces on two chips, a test controller can drive a series of values onto the outputs of one chip and sample the input values on the pins of another chip connected to those outputs.

The most popular interface to boundary-scan implementations is IEEE standard 1149.1—popularly known as JTAG (from "Joint Test Action Group," the name of the committee that devised the standard). JTAG defines a four-wire serial interface to dedicated on-chip test circuitry.

With the addition of special on-chip circuitry, a JTAG boundary-scan interface can also be used to communicate with on-chip debugging circuitry. Interestingly, most DSP processors that support JTAG-based on-chip debugging do not actually support boundary scan. That is, the JTAG standard is simply used as a convenient mechanism for communicating with the on-chip debugging circuitry.

Microprocessors and DSPs increasingly feature scan-based debugging capabilities. DSPs using JTAG to access scan-based debugging include the Analog Devices ADSP-210xx family, AT&T's DSP16xx family, the NEC μPD7701x family, and the Texas Instruments TMS320C2xx, TMS320C3x, TMS320C4x, TMS320C5x, TMS320C54x, and TMS320C80 families. All Motorola DSPs except the DSP56000, DSP56001, and the newer DSP563xx family feature a scan-based debugging facility called OnCE (On-Chip Emulation) that uses a proprietary serial protocol that is similar to JTAG. (The DSP56000 and DSP56001 do not have on-chip emulation, and the DSP563xx family uses JTAG to access its on-chip emulation facilities.)

Capabilities of on-chip debugging hardware vary widely. At a minimum, all processors with on-chip debugging hardware allow the user to read and write processor registers and memory. Most provide one or two *hardware breakpoints* (also called *address match breakpoints*). These breakpoints are implemented by loading an address into the on-chip debugging hardware, as opposed to replacing an instruction in the processor's program with a software trap instruction to implement the breakpoint. An advantage of hardware breakpoints on some processors is that they can also be used to halt the processor when it attempts to access a data memory location (as opposed to a program memory location). The on-chip debugging hardware on some processors provides a *program discontinuity trace buffer*, which saves the program counter values for the last several times that the processor branched or responded to an interrupt. This historical information can be valuable in understanding why the processor wound up executing a particular piece of code.

DSP processor vendors do not usually describe the exact capabilities of a processor's on-chip emulation hardware in the processor's documentation. (Motorola is an exception, providing thorough documentation on their processor's "OnCE" port.) Instead, the user must typically infer what capabilities are present from looking at the capabilities provided by the processor's in-circuit emulation software tools. For more information on in-circuit emulation tools, please refer to the subsection on in-circuit emulators in Chapter 16, "Development Tools."

Chapter 12

Power Consumption and Management

The number of portable DSP applications has increased dramatically in recent years. DSPs are now commonly found in portable devices such as cellular telephones, pagers, personal digital assistants, laptop computers, and consumer audio gear. All of these applications are battery powered, and battery life is a key product differentiator in all of them. As a result, designers are constantly looking for ways to reduce power consumption.

DSP vendors have responded to this challenge in multiple ways. First, almost all DSP manufacturers have introduced low-voltage DSPs, capable of operation at nominal supply voltages of 3.0 or 3.3 V. Second, many vendors have added power management features that reduce power consumption under software or hardware control. Both of these approaches are discussed below.

12.1 Low-Voltage Operation

One of the first steps DSP vendors have taken to reduce power consumption is to reduce the required processor supply voltage. Virtually all of today's DSP processors are fabricated using CMOS technology. Because power consumption in complementary metal-oxide semiconductor (CMOS) circuitry is proportional to the square of voltage, considerable savings can be realized by reducing supply voltage. Most DSP manufacturers now offer DSPs that use a nominal 3.3 V supply (typically called *three-volt parts*), which reduces power consumption by about 56 percent relative to the same processor operating at 5.0 V. Some of the low-voltage DSPs (such as those from AT&T and some from Texas Instruments) can run with a nominal voltage as low as 3.0 V, further reducing power consumption.

Note that we used the word "nominal" above when discussing DSP processor supply voltages. Typically, electronic components such as DSP processors can function within a range (usually ±10 percent) of their nominal supply voltage. Thus, a 3.3 V nominal device can actually operate with a supply voltage from about 3.0 V to about 3.6 V. Similarly, a 3.0 V nominal device can operate with a supply voltage ranging from 2.7 to 3.3 V. In an attempt to gain a marketing advantage, some DSP processor vendors have taken to stating the *minimum* supply voltage when describing their processors. This can result in confusion: if a part is described as operating at

"3.0 V," does it mean that its nominal supply voltage is 3.0 V, or 3.3 V? For consistency, throughout this book we use nominal voltages in all cases.

In some cases, low-voltage DSPs are actually 5 V parts that are able to run at a lower voltage. In such cases the system clock must be reduced to permit operation at a lower voltage. DSP vendors have also begun to introduce "true" 3 V versions of their DSPs that are able to run at full speed at 3.0 or 3.3 V.

DSPs available at low voltages include the AT&T DSP16xx and DSP32xx families, the Analog Devices ADSP-2103, ADSP-2173, ADSP-2183, and ADSP-2106x; the DSP Group PineDSPCore and OakDSPCore; the IBM MDSP2780; the Motorola DSP56L002 and DSP563xx; the NEC μPD77015, μPD77017, and μPD77018; the SGS-Thomson D950-CORE; and the Texas Instruments TMS320LC3x, TMS320LC5x, TMS320VC54x, and TMS320C80.

12.2 Power Management Features

In addition to low-voltage operation, many processors incorporate *power management* features. These features provide control over the frequency of the processor's master clock or over which parts of the processor receive the clock signal. These approaches work for two reasons. First, CMOS circuitry consumes energy at a rate linearly proportional to its clock frequency. The higher the clock frequency, the more power the device consumes. Second, the more parts of the chip that are active (i.e., clocked), the more power the device consumes. Thus, reducing the clock rate or limiting the portions of the processor that are clocked lessens power consumption.

Sleep and Idle Modes

Sleep (or *idle*) modes typically turn off the processor's clock to all but certain sections of the processor to reduce power consumption. While asleep, the processor does no work; it simply sits idle, waiting for a wake-up event of some kind. In most cases, this event can be any unmasked interrupt. If the interrupt can come from on-chip peripherals, this implies that these peripherals are active and receiving a clock signal, increasing power consumption. As a result, some processors have two sleep modes: one in which on-chip peripherals are enabled (and can thus interrupt and wake the processor) and one in which on-chip peripherals are disabled. In the latter mode, only an external interrupt (which may have to come from one or two special pins) can wake the processor. In fact, some processors can only be awakened from certain sleep modes via the RESET pin.

Sleep modes are usually entered by executing a special IDLE (or similarly named) instruction, or by setting a bit in a control register. Some processors also provide pins that an external device can assert to force the processor into a power-down state. Processors that have power-down pins may also provide a power-down acknowledge pin that signals to the external circuitry that the processor has entered its sleep mode.

Wake-up latency may be a concern in some applications. Wake-up latency is the amount of time the processor requires to exit a sleep mode and resume normal execution. In cases where the processor uses an external clock as its master clock signal, this wake-up latency is usually one or two instruction cycles. As will be discussed in Chapter 13, a processor may use an internal oscillator in combination with an external crystal to generate a clock signal. In this case, if the sleep mode disables the internal oscillator then a wake-up latency of thousands (or even tens of thousands) of instruction cycles may be necessary while the oscillator stabilizes. A similar problem occurs with phase-locked loop (PLL) on-chip clock generators: turning off the PLL reduces power consumption, but the PLL may require a significant amount of time to lock when it is reenabled.

In a further effort to reduce power consumption, some processors (for example, the Analog Devices ADSP-2171 and the Texas Instruments TMS320LC31 and TMS320C32) provide slow-clock idle modes. These modes reduce power by slowing down the processor's master clock before going to sleep, but they usually also increase wake-up latency.

Clock Frequency Control

Many applications do not require the processor's full execution speed at all times. For example, a digital cellular telephone spends most of its time idle. Even when it is in use, it may need full processing power only when one party is speaking and not during periods of silence.

To take advantage of this observation, several DSPs allow the programmer some degree of software control over the DSP's master clock frequency. This represents a compromise between full-speed operation and a sleep mode, since the DSP is still executing and consuming power, albeit at a reduced rate.

The two most common clock control mechanisms are clock dividers and low-speed clock sources. A clock divider allows the system clock rate to be reduced by a programmable factor, which reduces execution speed and power consumption proportionally. For example, the Motorola DSP5600x family and the DSP56166 all provide four-bit clock dividers that allow the master clock rate to be divided by one of sixteen values, ranging from 1 to 65,536 in powers of 2. Similarly, DSP Group provides clock divider macrocells for their PineDSPCore and OakDSPCore DSP cores. The Texas Instruments TMS320LC31 and TMS320C32 have a LOPOWER instruction that reduces the clock frequency by a factor of 16.

Some DSPs (for example, some members of the AT&T DSP16xx family) can use a low-speed on-chip oscillator as a source of the master clock. While this is not as flexible as a clock divider that supports multiple clock speeds, it is sufficient for many applications.

Peripherals and I/O Control

As mentioned above, the more CMOS circuitry that is supplied with a clock signal, the more power a chip will consume. Because many applications do not make full use (or full simultaneous use) of all on-chip peripherals, almost all DSPs provide ways to disable unused peripherals.

Additionally, the analog sections of certain peripherals, such as on-chip A/D and D/A converters, may consume power even when their digital circuitry is not being clocked. To combat this, some DSPs have special power-down modes for their analog circuitry. DSPs with this capability include the Analog Devices ADSP-21msp5x processors and the Motorola DSP561xx family.

Disabling Unused Clock Output Pins

A number of DSP chips provide a clock output pin. This pin provides a square wave at the processor's master clock frequency (or a frequency closely related to it, e.g., one-half the master clock frequency) and is intended to provide a clock signal synchronized to the processor's master clock for use by external circuitry.

Some DSP processors can selectively disable this output pin. If the generated clock signal is not used, or is used only occasionally, disabling the clock output line reduces power consumption; the amount of power saved depends on the external capacitive load on the clock pin. Additionally, disabling the output clock also reduces electromagnetic interference generated by the DSP.

The Motorola DSP5600x and DSP56166, and the AT&T DSP16xx family are examples of processors that can disable their clock output lines.

Chapter 13

Clocking

DSP processors, like all synchronous electronic systems, use *clock signals* to sequence their operations. A clock signal is a square wave at some known frequency; the rising and falling edges of the waveform are used to synchronize operations within the processor.

The highest frequency clock signal found on a processor is called the *master clock*. Typical DSP processor master clock frequencies range from 10 to 100 MHz. Usually, all other lower-frequency clock signals used in the processor are derived from the master clock. In some processors, the frequency of the master clock may be the same as the instruction execution rate of the processor; such processors are said to have a *1X clock*. On other processors, the master clock frequency may be two or four times higher than the instruction execution rate. That is, multiple clock cycles are required for each instruction cycle. These processors are said to have a *2X* or *4X clock*. This illustrates the point that clock rates are not equal to instruction execution (MIPS) rates. To avoid confusion, we generally mention clock rates in this book only when discussing input clock requirements and use MIPS in all other circumstances.

A processor's master clock typically comes from either an externally supplied clock signal or from an external crystal. If an external crystal is used, the DSP processor must supply an on-chip oscillator to make the crystal oscillate. DSP processors that do not have on-chip oscillators must use an externally-generated clock signal. If an external clock at the right frequency is available, this is not a problem. However, if an appropriate clock signal is not available and your options are creating an externally-generated clock signal and using an external crystal with an on-chip oscillator, the crystal/oscillator approach is usually cheaper and saves both board space and power.

A number of DSP processors now have on-chip frequency synthesizers (also called *phase-locked loops* or *PLLs*) that produce a full-speed master clock from a lower-frequency input clock signal. On some processors, such as the Analog Devices ADSP-2171 and some members of the Texas Instruments TMS320C5x family, the input frequency must be one-half of the desired master clock frequency; these chips are said to have on-chip *clock doublers*. Other processors are slightly more flexible in the input frequencies they can handle. For example, the Texas Instruments TMS320C541 PLL can generate a full-speed clock from an input clock that is 1/3, 1/2, 2/3, 1, or 2 times the desired instruction execution rate.

A few processors provide extremely flexible clock generators. For example, on some members of the Motorola DSP5600x family and on the Motorola DSP561xx, the frequency synthesizer can generate a full-speed master clock from a very-low-frequency input clock. As an example, the Motorola DSP56002's frequency synthesizer can generate a 40 MHz master clock from a roughly 10 kHz input clock.

On-chip frequency synthesizers not only simplify designs and potentially reduce costs, but they can also reduce electromagnetic interference generated by high-speed external clock signals.

Chapter 14

Price and Packaging

Depending on the application, the importance of a DSP processor's cost may range from moderately important to critically important. In some applications, the DSP may represent only a tiny fraction of the overall system cost; in others, the DSP may comprise a large portion of the system cost. Some applications, such as those in consumer electronics and the personal computer industry, bring intense market pressure to cut costs. Product shipment volumes in such applications may be high enough that even minor differences in processor price make a significant difference in product profitability.

In this chapter, we examine pricing for several processors and discuss the types of IC packages that are frequently used for DSP processors.

14.1 Example Prices

Table 14-1 shows unit prices for a quantity 1,000 purchase of some of the processors covered in this book. The prices shown were current as of June, 1995. The processors chosen were the least-expensive versions of the fastest family members available.

From the table one can see that DSP prices vary from as low as $6 (for the TMS320C16 from Texas Instruments) to $450 (for the TMS320C80 multiprocessor DSP, also from Texas Instruments). Instruction execution rates range from 8.8 MIPS (again, the TMS320C16) to 70 MIPS (the AT&T DSP1627). Both performance and price cover very wide ranges indeed. In general, floating-point processors are more expensive than fixed-point ones, although the low price tag of the Texas Instruments TMS320C31 has allowed it to make inroads into applications previously reserved for fixed-point processors.

Several items should be noted about the prices shown in Table 14-1. First, the prices shown are the unit prices when purchased in a quantity of 1,000 units. Prices for larger volumes (e.g., 100,000 or 1,000,000 units) may be *significantly* cheaper. These high-volume prices are invariably set through negotiations with the processor vendor and are usually not publicly available, which is why they are not reported here. Second, DSP prices fluctuate and generally decrease over time. Thus, the prices shown in the table are most useful for relative comparisons. Third, other family members, package types, and speed grades may be available in addition to

TABLE 14-1. Representative Unit Prices of DSPs (as of June, 1995) Purchased in Quantities of 1,000. Package Nomenclature Is Explained in the Text

Vendor	Processor	Speed (MIPS)	Price (Qty. 1,000)	Pin Count and Package
Analog Devices	ADSP-2171	33.3	$38.00	128 PQFP
	ADSP-21062	40.0	$249.00	240 PQFP
AT&T	DSP1627	70.0	$49.80	100 PQFP
	DSP3207	20.0	$60.00	132 PQFP
Motorola	DSP56002	40.0	$38.20	144 TQFP
	DSP56166	30.0	$35.89	112 TQFP
	DSP96002	20.0	$156.90	223 PGA
NEC	μPD77015	33.3	$17.23	100 TQFP
Texas Instruments	TMS320C16	8.8	$6.06	64 PQFP
	TMS320C25	12.5	$12.30	80 PQFP
	TMS320C209	28.6	$14.00	80 TQFP
	TMS320C31	25.0	$54.10	132 PQFP
	TMS320C44	30.0	$158.40	304 PQFP
	TMS320C52	50.0	$22.80	100 PQFP
	TMS320C541	50.0	$43.95	100 TQFP
	TMS320C80	50.0	$450.00	305 PGA
Zoran	ZR38001	33.3	$42.00	128 PQFP

those listed in the table. The prices of these other versions may be considerably different from the prices listed above.

14.2 Packaging

Integrated circuits can be mounted in a variety of packages. Package options for some DSPs are shown in Table 14-2. The package option chosen can have a strong impact on an integrated circuit's price. As an example, a Texas Instruments TMS320C30 processor in a PGA package is almost twice as expensive as the same processor in a PQFP package.

TABLE 14-2. Popular IC Packaging Options

Package	Description	Comments
BGA	Ball grid array	Similar to PGA, but for surface mount; expensive; used for high-pin-count devices
BQFP	Bumpered quad flat pack	PQFP with nubs ("bumpers") on corners
CQFP	Ceramic quad flat pack	Better heat dissipation than PQFP
DIP	Dual in-line package	Inexpensive; usable up to about 40 pins
MQFP	Metal quad flat pack	Better heat dissipation than PQFP
PGA	Pin grid array	Usually most expensive; not surface mount
PLCC	Plastic leaded chip carrier	Inexpensive; usable up to about 68 pins
PQFP	Plastic quad flat pack	Inexpensive; most popular
SQFP	Shrunken quad flat pack	Another name for TQFP
TQFP	Thin quad flat pack	Thinner version of PQFP

In general, PQFP packaging is the cheapest for high pin-count devices, while PGA is the most expensive. CQFP devices are more expensive than their plastic counterparts, but are able to dissipate heat more effectively. PLCC and DIP packages are inexpensive, but are only suitable for devices with low pin counts.

Chapter 15

Fabrication Details

While it is often a secondary consideration, information about the fabrication process used to manufacture a DSP processor can provide useful insights for designers evaluating DSPs. The two most basic metrics that characterize fabrication processes for digital integrated circuits are *feature size* and *operating voltage*. In this chapter, we explore these issues and others relating to the fabrication process.

15.1 Feature Size

Feature size is used as an overall indicator of the density of an IC fabrication process. It usually refers to the minimum size of one particular kind of silicon structure or *feature*, specifically the minimum length of the *channel,* or active region of a MOS transistor. (MOS stands for metal-oxide semiconductor and refers to the type of transistor dominant in digital integrated circuits.) The sizes of other structures on the IC are usually roughly proportional to the minimum transistor channel length. The best widely used fabrication processes today have a feature size of 0.6 or 0.8 μm, and some manufacturers have 0.5 μm processes available. (A *micron* (μm) is one one-millionth of a meter.)

A smaller feature size translates into a smaller chip, which in turn translates into better performance and lower production costs. Lower production costs are possible because more chips can be manufactured using a single *wafer* of a fixed size, and yields are higher with a smaller die. (A wafer is the silicon disk on which chips are built. Multiple chips are fabricated on a wafer, then separated and placed into individual packages.) Better performance is made possible by the faster transistors, shorter interconnections, and lower capacitance that comes with reducing the size of devices on the chip.

Of course, the performance and selling price of a processor are specified by the manufacturer, so the system designer does not need to know the details of the fabrication process to determine these. Manufacturers often create improved versions of existing processors by converting to a smaller fabrication process. By analyzing the performance and cost of a processor that uses a particular fabrication process and considering the state of the art in fabrication technology, one can obtain a general feeling for the prospects for faster and cheaper versions of that processor. For example, if a given processor is only moderately fast, but is fabricated in an older, less aggressive process, then there is a good chance that substantial improvements in the processor's performance can be made by moving it to a smaller fabrication process if one is available to the manufacturer.

15.2 Operating Voltage

Until recently, virtually all DSP processors used an operating voltage of 5 V, which is the most common supply voltage for digital circuits. In the past few years, IC manufacturers in general, and DSP processor manufacturers in particular, have begun to produce components that operate at supply voltages of around 3 V. Because the power consumption of MOS circuits is roughly proportional to the *square* of the operating voltage, reducing the supply voltage from 5.0 to 3.3 V results in a reduction in power consumption of roughly 56 percent. Particularly in small, battery-powered devices like hand-held cellular telephones, this reduction in power consumption makes a dramatic difference for product designers. Reduced power consumption in the DSP processor translates directly into increased operating time with a given battery, since the DSP is usually one of the most significant consumers of power in these systems. Operating voltage and power consumption considerations are discussed in more detail in Chapter 12.

15.3 Die Size

Another physical characteristic of interest is a processor's *die size*. The die size is the size of the actual silicon chip containing the processor's circuitry. Chips with larger die sizes are more expensive to manufacture and may require larger, more expensive packages.

Die sizes are usually specified in square millimeters (mm^2), although for historic reasons they are sometimes quoted in square *mils*. A mil is one one-thousandth of an inch, or approximately 0.25 mm. Typical die sizes for DSP processors range from 7 by 7 mm to 15 by 15 mm.

For a DSP core that is used as a building block in a custom ASIC, the die size of the core strongly influences the overall die size of the ASIC. Or, viewed another way, if the overall size of the ASIC must be limited (for example, to meet manufacturing cost targets), then the die size of the core dictates how much silicon area in the chip is left over for custom features. DSP cores range in size from 3.9 mm^2 (for the Clarkspur Design CD2400 in 0.8 μm CMOS) to 10 mm^2 (for the SGS-Thomson D950-CORE in 0.5 μm CMOS).

Chapter 16

Development Tools

No matter how powerful a processor may be, a system designer requires good development tools to efficiently tap the processor's capabilities. Because the quality and sophistication of development tools for DSP processors varies significantly among processors, development tool considerations are an important part of choosing a DSP processor.

A detailed evaluation of the capabilities of DSP development tools is beyond the scope of this book. A separate report, *DSP Design Tools and Methodologies* [BDT95], provides in-depth analyses of these tools, related high-level design environments, and various design approaches. Here, we limit our discussion to reviewing the types of development tools critical to DSP processor-based product development and the key capabilities required of each type of tool. The material in this chapter is adapted from [BDT95].

16.1 Assembly Language Tools

Compared with other kinds of electronic products, DSP systems are often subject to extremely demanding production cost and performance requirements. Software for their programmable components often must be extremely efficient. Traditionally, DSP software has been written in assembly language, with heavy emphasis on optimization, and is usually developed from the ground up for each new system.

Such an approach incurs high development costs, as developing assembly language software for programmable DSPs is a time-consuming and difficult process. Compared to general-purpose processors, DSPs typically have very irregular instruction sets and highly specialized features that complicate programming. This is especially true of fixed-point DSPs. As in so many aspects of the design process, good tools can make a dramatic difference in the productivity of DSP software developers. In this section, we discuss tools for assembly language software development on DSP processors.

Assembly language development environments for DSP processor software may include assemblers, linkers, instruction set simulators, in-circuit emulators, debuggers, development boards, and assembly language libraries, as illustrated in Figure 16-1. These tools are usually specific to the processor being targeted. Often they are developed and distributed by the processor

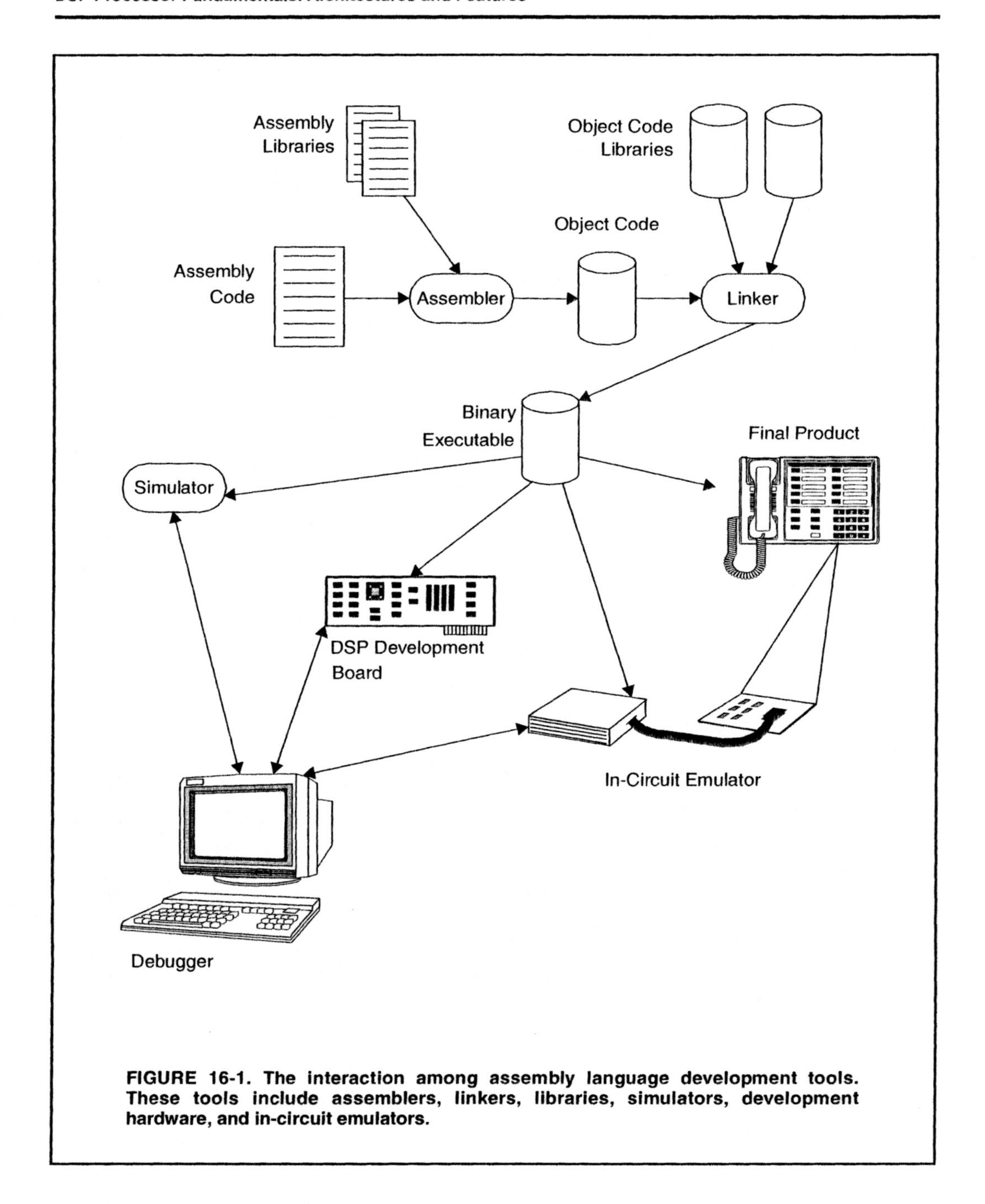

FIGURE 16-1. The interaction among assembly language development tools. These tools include assemblers, linkers, libraries, simulators, development hardware, and in-circuit emulators.

vendor; third parties also provide assembly language development tools. In fact, in some cases, DSP processor vendors rely on third-party firms to develop their assembly language tools rather than developing them in-house.

All DSP processor vendors provide a set of basic assembly language tools for each of their processors, although the quality and completeness of the tools vary. Texas Instruments' DSPs have historically had the most extensive selection of assembly language tools, including numerous third-party products. However, other processor vendors have recently begun to offer more sophisticated basic tools in some cases. For example, Analog Devices, AT&T, and Motorola have all demonstrated or are beta-testing Microsoft Windows-based DSP development tools with impressive features. One vendor, NEC, even offers an integrated development environment (IDE) that integrates a language-sensitive text editor and "makefile" utility along with the standard complement of assembler, linker, simulator, and debugger. The integration of these tools allows some very useful features, such as automatic highlighting of source code statements that cause assembler errors. Though IDEs have been offered for years for general-purpose microprocessors, they have only very recently begun to emerge for DSPs. This illustrates one of the stubborn paradoxes of DSP processors: good software tools for DSP processors are essential because of the need to optimize software, but DSP processor software tools generally lag behind their general-purpose processor counterparts in sophistication and quality.

It should be noted that assembly language tools are often also an integral part of higher-level tools. For example, assemblers, linkers, and assembly language libraries are often used by high-level language tools such as C compilers or block-diagram-based code generation systems.

Assemblers

An assembler translates processor-specific assembly language source code files (ASCII text) into binary object code files for a particular target processor. Usually this object code requires the additional steps of linking and relocation to transform it into a binary executable file that can be run on the target processor.

Assemblers are among the most basic DSP software development tools; they are also among the most important, since a vast amount of software for DSP processors is developed in assembly language to optimize code speed and size. Below, we describe the most important features of DSP processor assemblers.

Most assemblers for programmable DSP chips are *macro assemblers*, meaning that the programmer can define (or use predefined) parameterized blocks of code that will be expanded in-line in the final program. Macros are an important feature, since they allow the programmer to reduce the amount of source code that must be maintained, while avoiding the overhead of subroutine calls. Figure 16-2 illustrates the definition and use of a typical assembler macro.

Another useful assembler feature is *conditional assembly:* the ability to conditionally control whether certain blocks of code are assembled, depending on the value of an assembler meta-variable. This can be useful for maintaining multiple, slightly different versions of a single program without having to create multiple copies of the source code. For example, an audio encoding application might use conditional assembly based on the value of a variable to decide whether to assemble code for monaural or stereo processing.

```
; Define macro "FIR" to implement FIR filter.
; Parameter "ntaps" defines number of taps.
;
FIR    macro ntaps
       clr   a
       rep   #ntaps-1
       mac   x0,y0,a  x:(r0)+,x0  y:(r4)+,y0
       macr  x0,y0,a  (r0)-
       endm
```

(a)

```
; Set up coefficient and data pointers.
       move  #data,r0
       move  #coeffs,r4
; Call macro for 256 taps.
       FIR   256
```

(b)

FIGURE 16-2. (a) Macro definition for an FIR filter for the Motorola DSP5600x. (b) Example usage of the macro.

COFF (*common object file format*) is a standard format for object code files and is supported by many assemblers. COFF allows the annotation of object files with debugging information, such as pointers to the lines of source code that correspond to each machine instruction. The use of COFF can also simplify the integration of third-party tools (such as debuggers) with a processor vendor's assembly language development tools.

For more details on assembler features and evaluation criteria, please consult *DSP Design Tools and Methodologies* [BDT95].

Linkers

Linkers are used to combine multiple object files and libraries into an executable program. To do this, linkers must *relocate* code; that is, fix the addresses at which various code fragments in object files and libraries are to reside when they are executed. With some processors, a linker may not be needed for small applications, but for most serious applications it is a necessity.

Because linkers are used to bring together many parts of an application program, they must be flexible enough to accommodate the requirements of different object files and memory configurations.

Most linkers for digital signal processors are based on the concept of a *memory map*, which tells the linker which segments of memory (and which memory spaces) to use for each section of program code and data. Because different system hardware designs and applications require different memory maps, the user must be able to specify these maps. Linkers provide various ways of describing the desired memory maps. This process is more straightforward with some linkers than with others.

When a vendor's assembler generates COFF object files, the linker must support this format as well, since the linker processes object files produced by the assembler.

When debugging or analyzing a program, users frequently need to display the addresses of symbols used in both object and executable files. This information is stored in a structure called a *symbol table*. Some linkers are able to produce an ASCII representation of the symbol table at the end of the linking process, while others rely on separate utility programs to do this.

In linker parlance, a *library* is a group of object code files bundled into a single file. Libraries are sometimes called "archives." Libraries are a convenient way to manage large groups of commonly used subroutines or functions while avoiding the need for a large number of separate files. Virtually all linkers allow programs to be linked with code extracted from libraries.

Users frequently want to produce their own libraries. This is typically done with a separate program, called a *librarian* or *archiver*, which bundles object files into a single library. Surprisingly, not all DSP processor vendors give their users the ability to create libraries.

Instruction Set Simulators

Instruction set simulators are programs that simulate execution of a processor at an instruction-accurate level. Instruction set simulators provide a software view of the processor: that is, they display program instructions, registers, memory, and flags, and allow the user to manipulate register and memory contents. They are a key tool for DSP software debugging and optimization in most applications. Many programs are simulated on an instruction set simulator before being tested on hardware, since the simulator may offer a more controlled, flexible, and interactive environment than a development board, emulator, or target system. Also, instruction set simulators are often available well in advance of actual silicon or target hardware, thus enabling software development to begin before sample chips are available. Accuracy concerns and relatively slow simulation speeds (both of which are discussed below) are the primary limitations of instruction set simulators.

Most, but not all, DSP processor vendors offer instruction set simulators for their processors. Some of the smaller vendors of DSP processor cores, such as Clarkspur Design and Tensleep Design, do not provide instruction set simulators for their DSPs. In addition, a few processors targeted exclusively at personal computer- and workstation-based multimedia applications, such as IBM's MDSP2780 ("Mwave"), do not have instruction set simulators. For these applications, a stand-alone DSP simulator is less useful, because generally the DSP works in close cooperation with the host processor. Nevertheless, the lack of an instruction set simulator can be a hindrance when chip samples are not available.

There are wide differences in the capabilities of instruction set simulators for different DSPs. Because there is usually only one instruction set simulator available for a particular processor, the choice of a processor equates to the choice of an instruction set simulator. Users should therefore carefully consider their simulation needs before choosing a processor.

All instruction set simulators provide the user with the ability to single-step through a program and to view and edit the contents of registers and memory. The main factors differentiating instruction set simulators from one another are their accuracy, speed, completeness, and support for debugging and optimization. Each of these is discussed below.

- **Accuracy.** When evaluating an instruction set simulator, two types of accuracy are of interest: functional accuracy and timing accuracy.

  Functional accuracy refers to the fidelity with which a simulator models the functionality of the processor. That is, the likelihood that a certain sequence of instructions operating on a certain set of data will yield the same results on the simulator as on the actual chip. The functional accuracy of instruction set simulators is sometimes less than perfect. Therefore, after relying on an instruction set simulator for software verification, developers are well advised to recheck the functionality and timing of their application using an emulator or development board (discussed below). Unfortunately, the most common means of learning about inaccuracies in instruction set simulators is by discovering them oneself. Some vendors do make an effort to track these inaccuracies and inform users of them.

  Timing accuracy refers to the fidelity with which the simulator models the time relationships of certain operations within the chip. On many processors some instructions take more than one instruction cycle to execute, and instructions may take varying amounts of time to execute depending on factors such as pipeline interlocks or internal memory bus contention. Some instruction set simulators ignore these effects, while others track timing to the level of one instruction cycle or even a fraction of an instruction cycle. Generally speaking, a simulator with better timing resolution is preferable, but there is a trade-off between timing accuracy and simulator execution speed. Even when a simulator attempts to model time accurately, it may not always succeed.

- **Speed.** Naturally, instruction set simulators are much slower than the processors they simulate. A typical DSP processor instruction set simulator executes on the order of thousands of instructions per second on typical PCs and workstations. We have, however, found speed variations of up to a factor of 5 among different instruction set simulators.

- **Completeness.** Some instruction set simulators model only the core processor itself and ignore supporting hardware units such as on-chip timers, I/O interfaces, and peripherals, or simulate these functions very crudely. Others model the behavior of these functions quite accurately. If an instruction set simulator does not model these kinds of on-chip functions, it can be difficult to meaningfully simulate a complete application.

  However, some simulators provide interfaces to allow the user to add functionality to the simulator, and this can offset the lack of accurate modeling of some on-chip functions. Extensibility is discussed below.

Those simulators that do provide some modeling of I/O interfaces generally use files on the host system as the sources and destinations for data arriving at the processor's inputs and being sent from its outputs.

- **Debugging and optimization support.** The user generally interacts with an instruction set simulator through a front-end program called a debugger. Often, the same debugger can be used with the instruction set simulator and with the emulator for a given processor. Sometimes the same debugger also works with development boards. The debugger provides much of the debugging and optimization functionality for a simulator or emulator. Thus, the quality of the debugger has a strong impact on the usefulness of an instruction set simulator. Debuggers are discussed below. Note that when a processor vendor provides a high-level language compiler for a DSP, the simulator often provides support for debugging high-level language programs as well as assembly programs.

  Key debugging and optimization features provided by instruction set simulators include data display and editing, breakpoints, symbolic and source-level debugging, single-stepping, profiling, and cycle counting. Of these, data display and editing, single-stepping, and symbolic and source-level debugging are primarily features of the debugger rather than the simulator. As such, these features are discussed below in the section on debuggers. Breakpoints, profiling, and cycle counting are discussed briefly here.

  All instruction set simulators provide *breakpoints*: user-specified conditions which cause the simulation to be halted. Some simulators can handle only simple breakpoints, which cause the simulator to halt when a particular program location is reached. Others provide very sophisticated breakpoints, which can be conditioned on a complex arithmetic expression involving memory and register contents. This type of capability can be extremely useful for understanding obscure program bugs.

  Profiling support helps the developer ascertain the amount of time a program spends in different sections of code. Profiling data are usually collected either on a per-instruction basis or on a per-region basis. In per-instruction profiling, the debugger keeps tabs on the number of executions of each instruction in the program. In per-region profiling, the debugger tracks the amount of time spent in various regions of the program. These regions may be defined to correspond to functions or subroutines in the user's program. As most DSP applications are extremely performance-sensitive, profiling is an important capability. Perhaps surprisingly, good profiling support is found in only a few instruction set simulators, such as those from Texas Instruments and AT&T.

  Cycle counting allows the user to obtain a count of the number of instruction cycles executed between any two points in a program. This can be quite useful for determining the performance of an algorithm, as well as for determining average and/or worst-case execution time of a subroutine. Cycle counting is especially handy for understanding program behavior when complex pipeline dependencies cause some instructions to require extra cycles for execution.

- **Integration/extensibility.** Instruction set simulators are generally designed to simulate the DSP processor alone and do not provide capabilities for simulating the behavior of peripherals or other processors. (There are some exceptions: those from DSP Group and Motorola provide the user the ability to link in custom C code to simulate external devices; Motorola also provides built-in support for simulating multiprocessor configurations.) Most instruction set simulators do provide memory-mapped file I/O, meaning that data words written to or read from a particular simulated memory location can be mapped to a file on the host computer. Some simulators provide port- or pin-level I/O, meaning that various processor I/O interfaces can be mapped to a file, and pin values (0, 1, or "undefined") can be read from or written to the file. Some simulators provide scripting languages that can be used to specify input data sequences and interrupt arrival times for the processor's I/O interfaces. For example, NEC's simulator for the μPD7701x has a particularly powerful scripting language for simulating I/O activity.

Simulation Models for Hardware/Software Cosimulation

Because of the increasing integration between hardware and software and the emergence of new design techniques such as DSP core-based ASICs and customizable DSPs, developers of DSP systems often need more sophisticated simulation models than the instruction set models traditionally provided to support DSP processor software development. The following paragraphs briefly describe the hardware/software cosimulation model options available to designers. However, the details of these models and the issues surrounding their use are beyond the scope of this book. We urge readers interested in this area to consult *DSP Design Tools and Methodologies* [BDT95].

The most useful hardware/software cosimulation model is the so-called *full-functional model*. This can be thought of as an instruction set simulator that can be embedded in a larger hardware or system simulation and that accurately models the processor's I/O pins as the model simulates processor execution. This type of model can be extremely useful for hardware/software cosimulation if it also provides a debugging interface that allows the user to issue standard debugging commands (e.g., examine and modify registers and memory, set breakpoints, and so on).

Alternatives to full-functional models include bus-functional models (which mimic the processor's I/O pins but do not actually simulate the execution of code), hardware modelers (which employ a physical sample of the processor to provide both hardware and software simulation), and design-based models (which use a low-level description of the processor generated to provide a very accurate, but very slow, simulation of both hardware and software).

In-Circuit Emulation

In-circuit emulation (ICE) assists the user in debugging and optimizing applications running on the DSP in the target system. External hardware and software on a personal computer or workstation provide the user with the ability to monitor and control the processor in the target system as it executes application programs.

All emulators provide the user with the ability to single-step through a program, and to view and edit the contents of registers and memory. The main factors differentiating emulators from one another are the maximum processor speed supported, the ability to trace program flow and pin activity in real-time, the sophistication of real-time and non-real-time breakpoint conditions supported, and profiling capabilities. Each of these is discussed briefly next.

- **Processor speed.** Key considerations for evaluating an emulator include whether or not it supports full-speed operation of the target processor and, if so, what set of debugging features are supported with real-time operation. These factors depend mostly on the architecture of the emulator. Emulator architectures are discussed in detail later in this subsection.
- **Program flow and pin tracing.** Some emulators provide high-speed buffers which can be used to trace the flow of program execution and in some cases to trace the activity on processor I/O pins as the processor executes in real-time. Real-time program trace is implemented by capturing the state of the processor's program memory address bus during each instruction cycle. Real-time program trace capability can be extremely useful, since program bugs that appear when the processor executes in real-time may disappear if the processor is halted or placed into a slow, single-stepping mode. Similarly, having the ability to capture traces of processor I/O pins in real-time can be quite helpful for debugging.

  Some emulators that do not provide true program flow tracing still provide limited information on program flow through the use of a *discontinuity buffer.* The discontinuity buffer typically records the source and destination addresses of the last few jump, call, or return instructions—or other instructions which cause the processor's program counter to change in a way other than being incremented.
- **Breakpoints.** *Real-time breakpoints* allow the processor to run at full speed until a specified program address is reached or another specified *trigger condition* is met. All emulators support real-time breakpoints that stop program execution at a specified program address. Other kinds of real-time breakpoint conditions supported by some emulators include accesses to a data memory location (possibly distinguishing between read and write accesses), accesses within a range of program memory locations or within a range of data memory locations, and execution of certain kinds of instructions such as program branches.

  Some emulators include one or more counters in their breakpoint logic. A counter allows the user to program the emulator to halt the processor after a particular condition has occurred a certain number of times. The Texas Instruments TMS320C4x and TMS320C5x emulators provide the most sophisticated real-time breakpoint capabilities among commercially available DSP processors.

  Non-real-time breakpoints are used if the needed breakpoint trigger condition is more complex than the emulator's breakpoint logic can support. Examples include halting execution upon the change of the value in a register or memory location, or when an arithmetic or logical expression involving two or more memory locations becomes true. Because such complex breakpoint expressions are beyond the capabilities of the emulator's breakpoint logic, they must be evaluated by the debugger after the execution of each instruction. Therefore, the processor is placed in a single-stepping mode and the emulator forces the DSP to execute debug instructions after each instruction. Some emulators do not provide any support for non-real-time breakpoints, while others provide very sophisticated support.
- **Profiling.** *Profiling* is used to determine where a program spends most of its time during execution. Basic profiling support provides the user with a statement of instruction execu-

tion counts on a particular block of code. From this information, the user can decide how to best optimize the program. Profiling is commonly supported in instruction set simulators, but among emulators for commercially available DSPs, only the Texas Instruments TMS320C4x and TMS320C5x support profiling. The emulator sets breakpoints at the start and end addresses of loops, subroutines, and functions and measures processor cycle counts between the breakpoints. Since the emulator must intervene for each profiling event encountered, execution speed is slowed, but is still faster than the instruction set simulator.

The features and capabilities of an in-circuit emulator are largely determined by the basic architecture of the emulator. Emulator architectures can be divided into three categories:

- **Pod-based emulation.** With pod-based emulation, the DSP is removed from the target system. Hardware attached to the host computer (called an *ICE adapter* or *pod*) contains a sample (sometimes a special version) of the processor to be emulated, with additional hardware for controlling it. A cable with a connector whose pin-out is identical to the DSP's connects the pod to the target system, replacing the target DSP. This is illustrated in Figure 16-3. Compared to the alternatives, pod-based emulators have some strong advantages, such as the ability to provide real-time traces of program execution and processor pin activity. However, pod-based emulators are expensive and can be troublesome, since replacing the processor in the target system with the emulator pod changes the electrical drive and loading characteristics of the circuit, and may cause electrical timing problems. Pod-based emulators were once the only type of in-circuit emulators available. Over the past few years, scan-based emulators (discussed below) have become prevalent. Only a few DSP processor vendors and third-party tool providers still offer pod-based emulators (for example, Analog Devices still offers pod-based emulators for the older members of the ADSP-21xx family). Because of their hardware complexity and the electrical loading effects mentioned above, pod-based emulators do not always support full-speed processor operation.

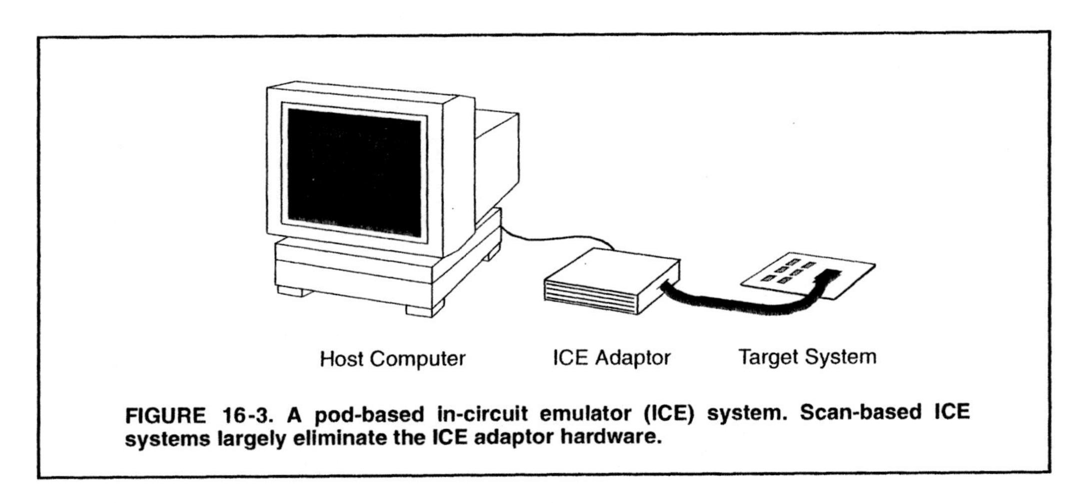

FIGURE 16-3. A pod-based in-circuit emulator (ICE) system. Scan-based ICE systems largely eliminate the ICE adaptor hardware.

- **Scan-based emulation.** Over the past five years or so, almost all DSP processor vendors have begun adding special debugging logic to their DSPs, along with a special serial port to access this logic. Some processors use a "JTAG" (IEEE standard 1149.1) compatible serial port to access these features. On other processors, a special dedicated interface is provided (e.g., Motorola's "OnCE" port). By connecting an IBM PC or workstation to the processor's serial debugging port (via a bus adaptor of some sort), the debugging features of the processor can be accessed. We call this approach *scan-based emulation.* With scan-based emulation, the on-chip debugging logic is responsible for monitoring the chip's real-time operation and halting the processor when a breakpoint is reached. After the processor is halted, the software communicates with the debugging logic over the serial debugging port.

  Scan-based emulation has several advantages over the traditional pod-based approach. First, the processor does not have to be removed from the target system and replaced with an emulator pod—a key consideration when processors cannot be mounted in sockets due to space or electrical constraints. Second, the number of signal lines that must be connected to the target hardware is minimized (typically the debug port has only five signals), and the debugging port signals do not have to operate at the same speed as the rest of the chip's signals. This reduces the overall complexity and cost of the emulator system and virtually eliminates the possibility of emulator attachment causing serious changes in the target system's behavior. Because the debugging logic is on-chip, scan-based emulators always support full-speed processor operation. However, scan-based emulators must revert to a very slow single-stepping mode to implement certain features (such as program flow tracing) which can be implemented in real-time with pod-based emulators.

  Because the debugging logic required to support scan-based emulation must be present on every copy of the DSP processor manufactured, its cost must be carefully constrained. For this reason, scan-based emulators generally have more limited capabilities than pod-based emulators. The capabilities of scan-based emulation are further limited by the serial connection between the target processor and the host, which has extremely limited bandwidth. One key feature found in pod-based emulators but not in scan-based emulators is real-time program flow and pin tracing.

  Scan-based emulation is presently supported by most DSP processor vendors, especially for processors introduced within the past few years. In addition, many third-party suppliers offer scan-based emulators for DSPs from Texas Instruments and Motorola.

- **Monitor-based emulation.** For processors which lack pod-based or scan-based emulators, or for applications where the use of these tools is inconvenient or prohibitively expensive, it is sometimes possible to obtain some of the functionality of an emulator by running a special supervisory program (called a *monitor*) on the DSP processor. One of the processor's conventional I/O interfaces (such as a host port) is used for communication with the debugger program running on the host. For example, monitor-based emulation is used with IBM's MDSP2780 Mwave DSP, which is intended for use in personal computer applications. In these applications, the monitor program communicates with the host processor through the same interfaces used by application programs. The key advantages of monitor-based emulation are that no special emulation hardware is required—either on the

DSP processor or externally—and that the processor does not have to be removed from the target system and replaced with an emulator pod. However, the debugging capabilities provided by monitor-based emulation are typically more limited than those found in pod- or scan-based emulation. For example, with monitor-based debugging it is usually not possible to set real-time breakpoints which are triggered by accesses to data memory locations, since there is no means provided for detecting such accesses. With monitor-based emulation, real-time breakpoints are typically limited to program memory locations. These breakpoints are implemented by substituting a call to the monitor for the instruction at the breakpoint address. After execution of the user's program is halted, the original instruction is replaced. Obviously, this approach cannot be used if the program code is executed out of read-only memory. Monitor-based emulation generally supports full-speed processor operation, but monitor-based emulators generally must revert to a slow single-stepping mode to implement some features (like program flow tracing) which pod-based emulators can implement in real-time.

A further disadvantage of monitor-based debugging is that when the monitor program is called (e.g., at a program breakpoint or when single-stepping), the state of the processor's pipeline is changed before it can be examined by the user. Despite its disadvantages, monitor-based debugging can be attractive because it is often very inexpensive and may be quite convenient to implement, especially in applications where the DSP is already coupled with a host processor.

Monitor-based debugging is primarily used in host-based applications such as personal computer multimedia and with low-cost development boards or evaluation boards. For example, GO DSP sells an $89 monitor-based debugger that works in conjunction with Texas Instruments' $99 DSP Starter Kit.

As with instruction set simulators, in-circuit emulators are used with front-end programs called debuggers. The debugger provides the user interface and much of the functionality of the emulator. Debuggers are discussed in more detail in the next subsection.

Debuggers

Debugger is the term we use to describe the front-end program that provides the user interface and much of the functionality of an emulator or instruction set simulator. A powerful debugger is a critical tool for software development. Emulators furnished by the processor vendor generally share the same debugger as the vendor's instruction set simulator. This is convenient because users do not need to learn different interfaces for the two tools. Emulators from third parties sometimes provide their own debugger or sometimes license the interface from the processor vendor (as is the case with many third-party emulators for Texas Instruments DSPs).

Debugger user interfaces vary significantly in their power and ease of use. The main types are:

- **Character-based, command-line oriented.** Command-line-oriented debuggers provide text-based interaction within a single window, much like MS-DOS commands and their output. While simple and straightforward, such user interfaces can make it difficult to debug complex applications. AT&T's debuggers for their floating-point DSPs use this approach.

- **Character-based, windowed.** A number of vendors provide debuggers which are character-based, but which divide the screen into windows for displaying different types of information. These are usually an improvement over command-line-oriented types. Texas Instruments and Analog Devices use this approach for most of their debuggers. Figure 16-4 shows a view of Texas Instruments' TMS320C3x debugger.
- **Graphical, windowed.** An increasing number of debuggers provide true graphical windowing by taking advantage of common windowing systems such as Microsoft Windows or the X Window System. This type of user interface is usually the most powerful, flexible, and easy to use, allowing extensive customization of the screen layout to suit the application and the ability to share the screen with other applications. Examples of debuggers using this approach include those from DSP Group (for the PineDSPCore and OakDSPCore) and NEC's debugger for the μPD7701x. NEC's debugger is shown in Figure 16-5. Analog Devices, AT&T, and Motorola have all demonstrated or are beta-testing similar debuggers for their processors.

Beyond the type of user interface, the most important features distinguishing one debugger from another are support for symbolic and source-level debugging, on-line help for debugger commands and for the processor itself, sophisticated data display and entry capabilities, support for "watch variables," in-line assembly and disassembly, and command logging and scripting. These features are discussed below.

- **Symbolic debugging.** Refers to the ability to manipulate objects (such as variables) in the program being debugged by using their symbolic names, as opposed to addresses or other numeric values. Note that symbolic debugging is important both for assembly language and high-level language programming. Almost all debuggers provide basic symbolic debugging support.

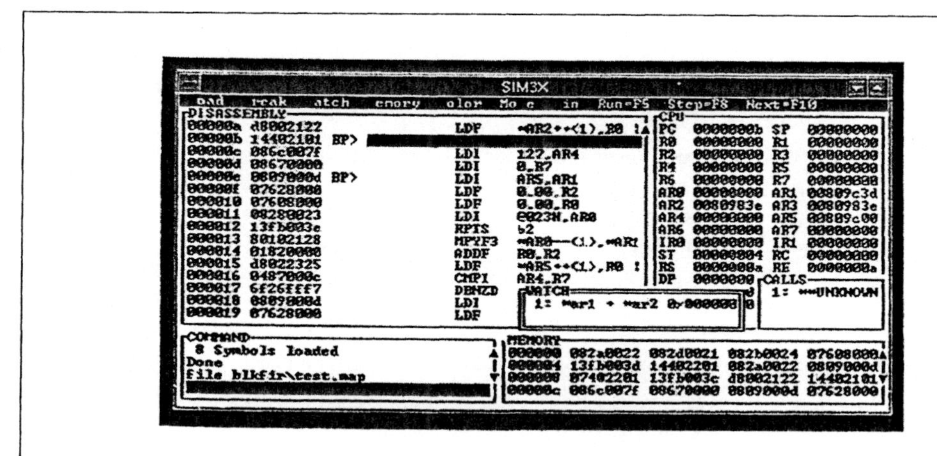

FIGURE 16-4. Texas Instrument's TMS320C3x debugger is typical of windowed, character-based debuggers.

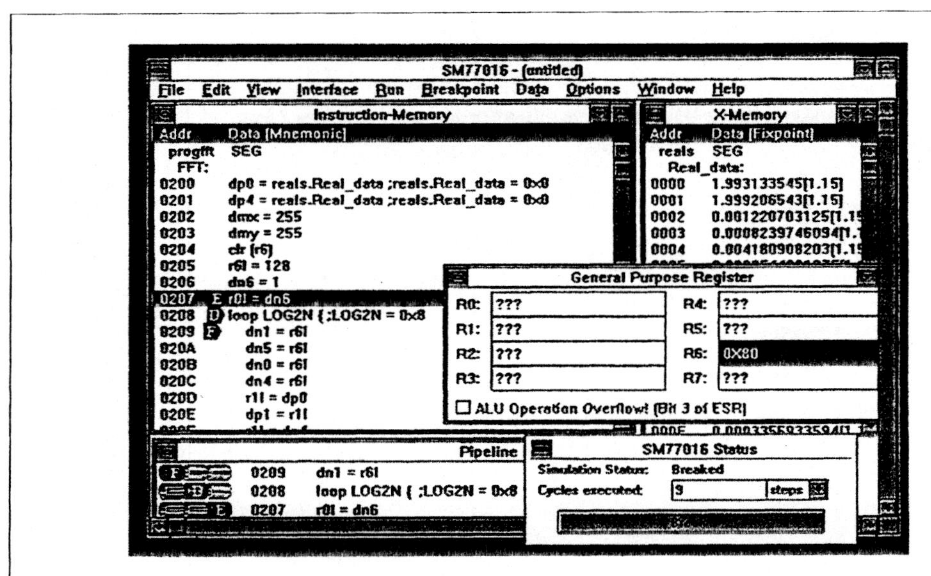

FIGURE 16-5. NEC's debugger for the μPD7701x provides a flexible, graphical user interface under Microsoft Windows.

- **Source-level debugging.** Refers to the ability to manipulate objects in the application program by referring directly to the source code that produced them. Typically, a window displays the source code (either assembly language or a high-level language, such as C) for a program and highlights each line as the simulator single-steps through the program. Some debuggers are capable of source-level debugging for both C and assembly code, while others can handle only one or neither.
- **Data entry and display mechanisms.** These are key to efficient debugging. It is of the utmost importance that debuggers allow the user to focus on the important aspects of a problem, and flexible data display is one of the best ways to achieve this. For example, some debuggers visually highlight data that have changed from one point to another in a program. Some debuggers provide the ability to display data using a selection of formats such as hexadecimal, decimal integer, and decimal floating-point; some can display data only in hex, which can be a serious inconvenience. Debuggers should also provide good facilities for displaying profiling data collected by the simulator or emulator, but very few do. Zilog's Z893xx debugger is one that provides good capabilities in this area.
- **Signal plotting.** This capability found in some debuggers, such as those for IBM's MDSP2780 Mwave DSP, allows the user to graphically display signals. This can be a very convenient feature, although good file export capabilities coupled with third-party graphing tools can also be used to good effect.

- **Watch variables.** Sometimes called *watch windows*, these allow the user to specify a set of registers or memory locations (or in some cases expressions based on register or memory contents) to be displayed in a special window. The variables displayed in a watch window are typically updated each time the simulator or emulator stops execution, e.g., after a breakpoint or single-step command.
- **Disassembly.** Recreates an assembly language program from a machine language executable. Note that this is different than source-level debugging: source-level debugging uses the original source file, while disassembly recreates an assembly language display from object code. In addition, an *in-line assembly* mechanism in some emulators and simulators allows the user to assemble instructions and store them in memory without having to reassemble and reload the entire program. This can be useful for quickly patching an errant program.
- **Command logging.** Provides the ability to create a log file of all commands executed during a particular debugging sequence. *Session logging* records not only the commands entered, but the debugger's output as well. *Scripting* or *macro capability* refers to the ability to execute commands from a file (which may have been generated from a previously recorded command log file). Because debugging sessions are frequently repetitive and require customization for maximum efficiency, scripting and logging facilities are important debugger capabilities.

Note that when a processor vendor provides a high-level language compiler for a DSP, often the debugger supports debugging high-level language programs as well as assembly programs. If the debugger does not provide this support, then debugging high-level language programs is extremely awkward. In some cases, third-party vendors of high-level language compilers provide separate source-level debuggers to accompany their compilers. For example, Intermetrics offers their XDB debugger in conjunction with their C compiler for Motorola's DSP96002.

Assembly Code Libraries

Optimized assembly code software libraries can be an extremely valuable tool for shortening software development time and improving software quality. Libraries are useful in two ways. First, in many cases a DSP software developer may be able to reuse optimized blocks from a library without modification. If the library functions are well tested and documented, this can result in a huge time savings. Second, even in cases where the software developer cannot find a needed block in the library, the existing library blocks may be very useful as examples of efficient coding style. In some cases, the software developer can create the needed function by making minor modifications to an existing library block. Of course, this is possible only if source code is provided. Some libraries are provided only in object code form, in which case they are sometimes referred to as *object libraries*.

We classify software libraries into two categories: *function libraries* and *application libraries*. Function libraries consist of relatively small building block functions, such as matrix multiplication, filtering, and I/O drivers. Libraries of these functions are often provided by DSP processor vendors, and listings may even be included in the processor databook. For example, Texas Instruments provides an extensive collection of function libraries for their programmable

DSPs, though they are in a disorganized state on Texas Instruments' electronic bulletin board system. More extensive, optimized libraries are typically provided by independent software vendors.

Application libraries contain much larger blocks of code that implement complete applications or parts of applications. Examples of this type of library include complete speech coders or modems. These libraries are usually developed and licensed by independent third-party software developers. For example, Analogical Systems provides speech coding and telecommunications libraries for the Motorola DSP5600x and Analog Devices ADSP-21xx DSPs. DSP Software Engineering provides a wide range of telecommunications applications for Texas Instruments DSP processors. In addition, some DSP processor vendors provide application libraries either free of charge or under license. A large number of assembly code library vendors and their products are described in *DSP Design Tools and Methodologies* [BDT95].

Note that libraries written in assembly code are often usable with users' C code as well as assembly code.

16.2 High-Level Language Development Tools

As mentioned previously, most DSP applications are typically programmed in hand-written assembly code to meet efficiency demands. However, the use of a high-level language (such as C) for software development has several advantages over assembly language:

- **Productivity.** It is much easier to develop error-free code in a high-level language than in assembly language.
- **Maintainability.** Because it is easier to understand than assembly language, high-level language code is generally easier to maintain.
- **Portability.** High-level language code is easier to move to a new processor than assembly language, which is, by definition, processor-specific.

The biggest drawback to the use of high-level languages is the loss of execution speed and increase in code size compared to hand-coded assembly language. Because DSP systems tend to be driven by demanding performance and cost requirements, such inefficiencies may not be tolerable in some systems. On the other hand, they may well be acceptable in certain applications, such as rapid prototyping and applications that do not require high sample rates or that have relatively modest computational requirements. Additionally, software development methodologies such as mixing assembly code and high-level language code can be used to increase the range of applications in which high-level languages can be used.

The following subsections discuss high-level languages available for DSP processor software development, and examine some of the advantages and disadvantages of using high-level languages.

The C Programming Language

C is the most popular high-level language for developing software for DSP processors. This is due to its familiarity to a wide range of programmers, its "closeness" to hardware (for example, its bit-wise logical operators), and its simplicity. Partially due to the features of the language, and partially due to the free availability of a good general-purpose C compiler (the GNU C compiler [GCC] from the Free Software Foundation), all currently available DSP processors and cores that have high-level language support have a C compiler. In fact, the only currently available DSP processor and core families that do *not* have C compilers available are the AT&T DSP16xx and the Zoran ZR3800x.

Despite its popularity, C lacks a number of essential language features that simplify coding of DSP algorithms. For example, C lacks both a fixed-point data type (a feature essential for efficient coding on fixed-point processors) and a complex data type (a feature that simplifies coding of DSP algorithms using complex arithmetic).

To address these issues, many compiler vendors have added their own extensions to the C language to improve both its suitability for DSP applications and its efficiency on DSP processors. For example, most C compilers support language extensions that allow the user to insert assembly language statements directly into the assembly code generated from the C program. Some also provide memory space qualifiers (i.e., keywords that inform the C compiler that certain variables are to be located in certain banks of memory).

In addition to ad hoc extensions used by compiler vendors, there are the efforts of the ANSI Numerical C Extension Group (NCEG), an ANSI committee working to extend the C language to better support numerical computations. At present, there is no official Numerical C standard; however, Analog Devices now offers a Numerical C compiler for their ADSP-210xx floating-point DSP processor. This compiler implements a subset of the enhancements discussed by the committee.

Other High-Level Languages for DSP Processors

Besides C, there are a number of other high-level languages that are used for programming DSP processors. The two most popular are C++ and Ada, which are described briefly below.

- **C++.** According to some DSP experts, C++ is the DSP high-level language of the future. A key advantage of C++ over C is that C++ developers can create new data types and operations to best address the needs of their applications. This increases the language's modeling efficiency: that is, fewer lines of code are needed to specify the desired processing. Additionally, the use of these data types and operators increases the compiler's opportunities for optimization.

  At present, only Tartan offers a C++ compiler for DSP processors. It is targeted at the Texas Instruments TMS320C3x and TMS320C4x DSPs.

- **Ada.** Ada offers a number of interesting features for DSP programming, especially for fixed-point processors. Ada is a more powerful language than C, supporting a wide range of constructs and data types. It is a "strongly typed" language with a rich set of types, subtypes, and type attributes.

  Ada's main strength—its expressive power—also results in its main drawback: the language is quite complex. For example, Ada has roughly twice as many keywords as C. Additionally, Ada is less suitable than C for some lower-level operations, such as bit manipulation.

Tartan is presently the only vendor offering an Ada compiler for DSP processors. The compiler is targeted at the Texas Instruments TMS320C3x and TMS320C4x DSPs. Tartan has also announced an Ada compiler for the Analog Devices ADSP-2106x family.

High-Level Language Efficiency Concerns

Historically, high-level language compilers for DSP processors have not met with great success. The biggest problem with high-level language compilers for DSP processors is that the code they generate is often not efficient enough (in terms of execution speed or memory usage) for use in production systems.

The main reason for this inefficiency is that DSP processors are, in general, extremely "compiler unfriendly." In order to reduce cost and increase performance, DSP processor vendors have spent years developing architectures that contain the minimal set of functions required to implement signal processing algorithms. In contrast, vendors of general-purpose processors have spent years attempting to make their processors *more* compiler friendly. Among the challenges modern DSP chips present to compiler writers are: multiple memory spaces, a small number of registers (typically of varying widths and dedicated to different uses), nonorthogonal instruction sets (meaning that certain instructions may only be legal with certain register or addressing modes), and no hardware support for software stacks. Additionally, the compiler must make good use of multioperation instructions (e.g., multiply-accumulate), parallel data moves, and hardware looping facilities in order to compete with hand-coded assembly language.

Another complication involves memory usage. Most DSP processors provide a small amount (typically several hundred to several thousand words) of on-chip memory. DSP processors typically execute faster when their instructions and data both reside in on-chip memory. Existing compilers tend to produce code that (in addition to being slower) is larger than its hand-coded equivalent, meaning that it more quickly consumes scarce on-chip memory.

These efficiency concerns are especially serious for fixed-point DSP processors for two reasons. First, fixed-point processors tend to be used in applications that are extremely cost sensitive. As a result, inefficiencies in execution time or memory usage are even more important than for floating-point processors. Second, the most popular high-level language for programming DSP processors—C—has no fixed-point or fractional data type. Emulating floating-point on a fixed-point processor is so expensive as to be out of the question for virtually all applications. Thus, unless the compiler vendor has extended the C language to include such a data type, programmers must make do with integer data and add explicit shifts to their code as necessary to implement desired fixed-point arithmetic. This complicates coding and lowers overall efficiency. Intermetrics' C compiler for the NEC µPD7701x and the Production Languages Corp. C compiler for Zilog Z893xx DSPs are examples of compilers that support fractional data types through language extensions.

Although compilers for floating-point DSPs also suffer from the inefficiencies outlined above, they have generally fared better than compilers for fixed-point processors. There are several reasons for this. First, floating-point DSPs tend to be used in less cost-sensitive applications. As a result, speed and memory efficiency are sometimes less of a concern. Second, floating-point DSPs tend to have wider instruction words and more registers, resulting in a more orthogonal instruction set. Third, such processors have hardware support for floating-point arithmetic, making them a good match for C's *float* data type.

High-Level Language Libraries

High-level language compilers usually include libraries of support functions callable from the high-level language. Vendors of DSP C compilers bundle a subset of the ANSI-standard C run-time library with their compilers. These libraries include standard utility functions used in many C programs, such as *strcpy, atoi,* and others. However, vendors often omit (potentially large) portions of the library—for example, standard I/O functions like *printf, scanf, getc,* and *putc* are often missing. Although ANSI-standard math functions (e.g., *sin, cos*) are usually included, they may not be particularly useful on fixed-point DSP processors. This is because these routines are specified by ANSI to use double-precision floating-point arithmetic, which must be (slowly) emulated in software on a fixed-point processor.

A large number of vendors provide function and application libraries callable from high-level languages. These include math and vector processing libraries (e.g., Tartan's VecTar library), libraries of signal processing functions (e.g., the DSP C Applications Library from Sonitech International), and applications libraries such as speech coders and modems. Some libraries are written in a high-level language, while others (for example, the DSP Expert Library for the TMS320C3x and TMS320C4x from Ariel Corp. and Tartan's FasTar math library for the same processors) are written in assembly code, but are callable from high-level languages. For more information on such libraries, please refer to *DSP Design Tools and Methodologies* [BDT95].

Debuggers

An important adjunct to high-level language compilers are source-level debuggers. It is unreasonable (not to mention unproductive) to provide a user with a high-level language compiler and then force him or her to use an assembly-level debugger to debug the resulting code.

In most cases, compiler vendors provide reasonable debugging support with their compilers. For example, Intermetrics offers an excellent source-level debugger for its Mwave C compiler. Similarly, Tartan provides the AdaScope debugger for its DSP Ada compiler.

16.3 Block-Diagram-Based Programming Tools

Many designers find block diagrams to be a very natural way to express signal processing algorithms. Block-diagram-based tools leverage this by allowing designers to design, simulate, or implement algorithms expressed as block diagrams. We refer to block-diagram-based tools that can generate programs that implement algorithms from block-diagram specifications as block-diagram-based programming tools.

Block-diagram-based programming tools can be used in several ways. First, they can provide a mechanism for rapidly generating C or assembly code to implement an algorithm. Second, they can provide a convenient way to access libraries, since libraries of blocks in a block-diagram-based programming tool correspond to function or application libraries in a conventional, textual programming language. Third, some block-diagram-based programming tools include simulation, fixed-point analysis, and hardware synthesis capabilities, as well as software generation, making them useful for many aspects of DSP system design.

The four most popular block-diagram-based programming tools are Alta Group's SPW, Mentor Graphics' DSP Station, Synopsys' COSSAP, and Hyperception's Hypersignal for Windows Block Diagram. Of these tools, the Hyperception software is the only one that runs on IBM PCs; the other three run only on UNIX workstations. All can generate C programs from block-diagram specifications. This C code can then be compiled, downloaded, and run on a DSP processor, although the generated C code is mainly intended for use on floating-point DSP processors. Additionally, SPW and DSP Station can directly generate assembly code for some processors (SPW supports the Motorola DSP5600x family, while DSP Station supports both the Motorola DSP5600x and the Texas Instruments TMS320C3x family).

For thorough analyses of these and other block-diagram-based programming tools, please refer to *DSP Design Tools and Methodologies* [BDT95].

16.4 Real-Time Operating Systems

An operating system is a program that controls access to system resources and manages the order of execution of separate programs (*processes*) on a processor. Real-time operating systems (or *kernels*) are operating systems that also provide guarantees about the real-time performance of the operating system. Typically, *real-time* means that the maximum amount of time between an interrupt occurring and a certain process being allowed to run is bounded. It may also mean that each process is guaranteed a certain percentage of the CPU, or that each is allowed to run within a specified time. Real-time operating systems can greatly simplify the problem of managing more than one task on a given processor.

It is important to note that use of a real-time operating system does not, by itself, guarantee that a system will meet its real-time constraints. In fact, it is easy to construct processes running under a real-time operating system that do not meet their real-time deadlines. Real-time operating systems must be viewed as tools that can be used to help the system designer construct a system with known real-time performance.

Real-time operating systems have long been popular for general-purpose microprocessors, such as the Motorola 68xxx family and RISC processors. A variety of real-time operating systems are now available on DSP processors as well. Some of these, such as SPOX from Spectron Microsystems, were designed for use on DSPs from the start. Others are DSP processor adaptations of real-time operating systems that were originally developed for general-purpose microprocessors. These latter systems are generally written in C (typically with a small amount of assembly language) and are typically available on floating-point DSPs due to their better C compiler support. Texas Instruments' TMS320C3x and TMS320C4x floating-point DSPs, for example, have a plethora of real-time operating systems available.

Fewer real-time operating systems are available on fixed-point DSP processors. Spectron Microsystems' SPOX is the leader in the fixed-point arena and is available on the Analog Devices ADSP-21xx, the Motorola DSP5600x and DSP561xx, and the Texas Instruments TMS320C5x fixed-point DSPs, as well as several floating-point DSPs.

For detailed evaluations of real-time operating systems for fixed- and floating-point DSPs, please refer to *DSP Design Tools and Methodologies* [BDT95].

16.5 Multimedia Environments

One of the advantages of a DSP processor over fixed-function hardware components is that the DSP processor can perform multiple functions, perhaps even simultaneously. An emerging application that exploits this advantage is the integration of DSP processors with personal computer and workstation platforms, where the DSP processor can be used as a fax or data modem, a music synthesizer, or a voice mail system, among many other possibilities. However, there is a cost associated with this flexibility. In particular, the execution of different tasks on the DSP processor must be choreographed in such a way as to ensure that the real-time requirements of each task are met. This capability is typically provided by a real-time operating system, as discussed above. In addition, if it is anticipated that multiple vendors will be providing DSP task software, then a standardized interface must be provided so that vendors can write DSP tasks that can communicate with the operating system, and so that developers of applications for the host (i.e., PC or workstation) processor can interface their programs to tasks running on the DSP in a consistent manner.

A *multimedia operating environment* includes a real-time operating system with a standardized interface for DSP functions, software that resides on the host processor to provide a means of communication between host applications and DSP tasks, and possibly also a library of DSP software tasks for such applications as modems and speech synthesis.

The first two multimedia operating environments to gain widespread attention were Mwave from IBM and Intermetrics, and VCOS (Visible Caching Operating System) from AT&T. AT&T's multimedia operating environment, VCOS, is no longer being actively marketed. WinSPOX is a relatively new, processor-independent multimedia operating environment from Spectron Microsystems. It is discussed in *DSP Design Tools and Methodologies* [BDT95].

Chapter 17

Applications Support

DSP processors are complex devices and can be tricky to design with, both from a hardware and a software standpoint. When choosing a processor, it is important to consider what kind of support the manufacturer provides to help ease the design process and solve problems as they arise.

Not only is there wide variability in the level of support provided by processor manufacturers, but there is also wide variability within a single processor manufacturer. Not surprisingly, high-volume customers (and potential customers) generally get better support than low-volume customers. However, some processor vendors (such as Analog Devices and Texas Instruments) have a strategy of providing broad support to all kinds of users, while other vendors (such as AT&T) concentrate on supporting a few major customers.

In the following sections, we outline the kinds of support typically provided by processor vendors and highlight the vendors who, in our assessment, provide the best support in each category.

17.1 Documentation

Documentation for DSP processors generally includes user's guides, data sheets, application notes, and application handbooks. User's guides provide an introduction to the architecture and the capabilities of the processor as well as a detailed discussion of the instruction set, memory architecture, peripherals, I/O interfaces, execution control, and instruction set encoding. Some user's guides include software benchmarks, example programs, and hardware designs using the processor. The average quality of DSP processor documentation is fairly poor, though there is quite a lot of variability among manufacturers. While errors of fact do find their way into these manuals, completeness and organization are usually more serious concerns. High-quality user's guides are a tremendous asset to system developers. Conversely, poorly organized or incomplete guides can be a major hindrance to efficient development. Analog Devices deserves special praise for their user's guides. They are clearly written, well organized, and provide several features that are surprisingly rare in DSP processor documentation, including a thorough index and a list of changes made in the guide since the previous version. AT&T's and Zoran's user's guides are relatively thin on details. Zilog's manuals use terminology that is at odds with accepted industry usage (e.g., referring to modulo addressing as "hardware looping") and contain large numbers of typographical errors. Those from Texas Instruments have tended to be massive but poorly organized and sometimes incomplete; this seems to be changing for the better.

Data sheets are usually used to give precise timing and pinout information for processor variations based on speed and packaging options. Some manufacturers include this information in their user's guides, rather than providing separate data sheets. As with user's guides, there is significant variation in the quality of data sheets among manufacturers. Accuracy is usually fairly good, but understandability is sometimes poor.

Application notes can be extremely useful in providing complete, practical hardware and software design examples. These examples can be a powerful tool for learning how to use a processor, for writing initial programs, and for understanding complex techniques for optimizing hardware and software. Motorola provides a good selection of application notes for the DSP5600x family. Application handbooks are compilations of application examples, emphasizing mostly software but including some hardware designs as well. Some handbooks include a diskette containing electronic versions of the examples described in the book. This is a great convenience. Again, Analog Devices is a standout here. The two-volume set *Digital Signal Processing Applications with the ADSP-2100 Family* [Ana90], [Ana95] documents a well-rounded collection of ADSP-21xx applications. The books include a tutorial discussion of each application as well as the source code and a discussion of the implementation used. A diskette containing the source code files is also provided. Texas Instruments also offers a four-volume set of application handbooks for their DSPs. These volumes consist mostly of reprints of technical articles originally published elsewhere. While they provide discussions of a good variety of applications, the Texas Instruments volumes lack the coherence and tutorial treatment of the Analog Devices books.

Many of the larger DSP processor manufacturers have *literature centers* that are responsible for distributing documentation. We include telephone numbers for vendors' literature centers in the Appendix, "Vendor Contact Information."

17.2 Applications Engineers

Applications engineers (sometimes called *AE*s) are probably the most important element of a DSP processor vendor's support infrastructure. These engineers are expert in the use of their company's processors. Among larger manufacturers, applications engineers are divided into those stationed in regional support offices around the world (*field applications engineers* or *FAE*s), and those stationed in the company's main engineering center.

The number and distribution of field offices varies widely from manufacturer to manufacturer. Some manufacturers provide all of their support from their central engineering center, while others have upwards of a dozen regional support offices worldwide.

Applications engineers usually participate in pre- and post-sales support. In pre-sales situations, they try to show the customer's design engineers why their processor is a better match for the application at hand than those offered by competing manufacturers. Applications engineers often get involved in detailed processor comparisons, sometimes even writing customer-specified benchmarks for their processors.

After a company has chosen a processor, the applications engineers can help designers understand the complexities of designing hardware and software around the processor. As we have observed, DSP processors in general are very tricky to program efficiently, have very irregular instruction sets, and in some cases lack good development tools and thorough documentation. Because of this, a local, knowledgeable applications engineer who can provide assistance during product development is an extremely valuable asset.

Although they are generally well qualified and very helpful, most manufacturers' applications engineers are in short supply. Not surprisingly, the largest customers (or potential customers) generally get the lion's share of the AEs' attention, though this varies from vendor to vendor.

In response to the perennial shortage of applications engineers' time, some DSP processor manufacturers are increasingly trying to make use of applications engineers employed by their distributors. While we have found distributors' applications engineers to be helpful sources of basic information, so far we have not found them to have the depth of technical knowledge of their counterparts who are employed by the processor manufacturers.

17.3 Telephone Support

In addition to in-person applications engineering support, the DSP processor manufacturers who target a broad market provide telephone hot-line support. These telephone lines are manned by applications engineers, and users can call in to have questions answered. A list of these numbers is included in the Appendix, "Vendor Contact Information." As with DSP processor manufacturers' field applications engineers, the engineers manning these telephone lines are generally too few to meet demand. In many cases, these engineers also act as applications engineers supporting specific design projects for internal or external customers. As a result, users calling these help lines generally wind up leaving their questions in a voice mail message. Our experience is that the timeliness and quality of responses varies significantly. On average, telephone support is fair. Some manufacturers try to prevent callers from contacting a specific support engineer and instead direct all calls to a central number. This can be problematic when users receive incomplete or unclear answers to their questions and need clarification but are unable to speak with the applications engineer who is familiar with their problem.

In addition to calling in with questions, in some cases users can submit questions by fax or through vendors' electronic bulletin board systems (discussed below) or via electronic mail. Among the major DSP manufacturers, Texas Instruments pioneered the use of Internet electronic mail as a vehicle for communicating with applications engineers. Motorola, Analog Devices, and AT&T now also provide electronic mail access to their support personnel.

17.4 Bulletin Boards

Texas Instruments, Analog Devices, AT&T, and Motorola operate their own electronic bulletin board systems (BBSs) for users of their DSP processor products. Users with access to a modem can dial in to these systems, which most often provide libraries of functions, application notes, and question-and-answer files. In some cases, the bulletin board systems can be used to leave questions for technical support staff members. We've found these BBSs to be helpful occasionally for locating useful pieces of software. However, many are poorly organized, making locating any particular item of interest difficult. In addition, many of these systems are undersized relative to the demand for them; thus, users may have to battle busy signals to get through. Finally, each bulletin board system has its own (usually cumbersome) user interface that users must be prepared to deal with.

Many DSP processor vendors now provide access to their bulletin board systems through the Internet. Texas Instruments deserves praise for an exceptionally useful World Wide Web site, containing not only DSP applications code but also well-organized, searchable tables of product information and press releases. The Texas Instruments Web site even contains PostScript versions of data sheets for new processors.

In some cases, independent institutions have (sometimes with the manufacturer's blessing, sometimes without their knowledge) copied portions of the contents of a manufacturer's bulletin board onto a computer that is connected to the Internet and available for public access. Users who have access to the Internet can retrieve files using the Internet File Transfer Protocol (FTP).

17.5 Training

Most DSP processor manufacturers provide formal training classes for users of their products. These are usually one to three days in length and focus on one processor family (for example, a vendor's fixed-point or floating-point processor family). Costs are typically between $500 and $1,500 per person. In the United States, vendors rotate training courses through different locations around the country.

17.6 Third-Party Support

In addition to the manufacturers themselves, a number of independent organizations provide various kinds of support for users of DSP processors. Not surprisingly, there is more third-party support activity for the more popular DSP processors. The major DSP processor manufacturers publish directories of these third-party vendors and their products. Texas Instruments' *TMS320 Software Cooperative* and *TMS320 Third-Party Support* guides are stand-outs in this area. See the Appendix, "Vendor Contact Information," for information on contacting DSP processor manufacturers to request their third-party directories.

Development Tool Vendors

While DSP processor manufacturers generally provide basic software and hardware development tools for their processors, there is a growing industry of independent tool vendors. These include providers of comprehensive, high-level simulation and code generation tools like the Alta Group of Cadence Design Systems, Inc., suppliers of emulators like DSP Research Inc., and software library developers like Wideband Computers Inc. Development tools are discussed in Chapter 16.

Independent tool vendors can be a very useful resource for system developers using DSP processors. First, many of the tools sold by these vendors provide real productivity gains for developers of DSP systems. Second, the tools themselves may encapsulate significant expertise about a given processor. For example, software function libraries that have been carefully optimized by expert programmers are not only useful as building blocks for creating application software, but also as tools for learning how to efficiently program a given processor. Of course, this assumes that the library code is clearly written, well documented, and available in source code form.

For a comprehensive listing of DSP development tool vendors and their products, including detailed evaluations of many products, see *DSP Design Tools and Methodologies* [BDT95].

Board Vendors

Dozens of companies manufacture off-the-shelf printed circuit boards for use in developing DSP processor-based systems. In some cases, these boards are intended to be used as a development aid—for example, to allow real-time evaluation and debugging of software while custom hardware is under development. In other cases, boards are designed to be incorporated into an end product. In either case, many boards provide interfaces to allow users to connect their own custom hardware to the board. The manufacturers of these boards can be useful sources of information on hardware design issues for DSP processor-based systems. In addition, some board vendors are also suppliers of software libraries and development tools such as debuggers and emulators.

Textbooks

Over the past few years, several textbooks have appeared that combine an introduction to digital signal processing with an introduction to one of the popular DSP processors. We recommend contacting DSP processor vendors for a current list of related books. Examples include *Real Time Digital Processing Applications with Motorola's DSP56000 Family* by Mohamed El-Sharkawy [Els90], and *Digital Signal Processing with C and the TMS320C30* by Rulph Chassaing [Cha92]. For complete citations of these texts and others, see the References and Bibliography.

Consultants

The rapid growth in the number of applications using DSP processors has spawned a cottage industry of independent consultants specializing in software and hardware for DSP processor-based systems. The larger DSP processor manufacturers maintain consultant directories that can be useful in locating someone with the right expertise for a given project. Be aware that the DSP processor manufacturers generally do not rigorously screen the consultants listed in their directory; on the contrary, they may be motivated to collect as large a list as possible for appearances' sake. As always, *caveat emptor.*

Design Houses

A number of independent companies specialize in complete development of DSP processor-based products. These so-called design houses generally specialize in one or more applications areas, such as telecommunications or motion control. DSP processor manufacturers can provide names of design houses that have experience designing with their processors.

Training

Independent companies and educational institutions from time to time provide short courses covering some of the more popular commercial DSP processors, sometimes in conjunction with an introduction to digital signal processing. These courses are usually advertised in major trade magazines or through direct mailings. DSP processor manufacturers may also be able to provide information on third-party training courses.

Chapter 18

Conclusions

In this chapter, we briefly review the broad themes of this book, summarize our findings, and provide a perspective on the history, the state of the art, and the future of DSP processors and applications.

18.1 Choosing a Processor Architecture

To the casual observer, all DSP processor architectures may appear to be very similar. As our in-depth examination has shown, this is not the case. While most DSP processors share certain basic features, the range of meaningful architectural and performance variations is substantial, as are differences in areas such as on-chip peripherals to support high levels of integration, system cost, energy efficiency, and ease of software and hardware development. It is our hope that by presenting and analyzing these differences in a consistent and clear manner, we have made it practical for readers to quickly understand the key differences among DSP processors and the practical implications of these differences.

Which processor architecture is best depends totally on the needs and constraints of the application. If time to market is critical, a floating-point DSP may be a good choice, as software development is simpler than for fixed-point devices. If cost is the primary consideration, very inexpensive DSPs are available, but they come with trade-offs such as limited performance and challenging software development. If power consumption is a key constraint, then a low-voltage DSP with an architecture specialized to the application or a DSP core-based ASIC may be the solution. In most applications, several of these factors are important, as are many others. The challenge is to identify the architecture that provides the best combination of characteristics for a given situation.

Different architectures clearly excel in different areas. One of the most important of these areas is execution speed. Even with roughly comparable instruction cycle times, application execution speed can vary significantly. But gaining insight into real differences in processor performance is a challenge. In general, performance measures of all kinds should be regarded with skepticism. Comparisons based on MIPS and MFLOPS (millions of floating-point operations per second) are particularly suspect, given the large differences in the amount of useful processing that different processors can accomplish with a single instruction or floating-point operation. While it is our firm belief that benchmarks, if carefully chosen and fairly implemented, provide invaluable insights into processor performance, benchmark results should not be accepted at face value. To draw meaningful conclusions from benchmarks, one must look beneath the surface to understand what the benchmarks really measure and apply the relevant results to the application at hand. (This approach is discussed in detail in the industry report *Buyer's Guide to DSP Processors* from Berkeley Design Technology, Inc.)

Additionally, designers should keep in mind that many features important to processor selection cannot readily be quantified via benchmarking. Applications support, quality of development tools, documentation, and I/O performance are all examples of processor selection criteria that are not captured in benchmark results. The careful designer must weigh both qualitative and quantitative considerations when choosing a processor.

18.2 DSP Processor Trends

More than fifteen years ago, the first DSP processors were developed by engineers whose primary expertise was in signal processing, not microprocessor technology. These processors evolved from fixed-function signal processing ASICs rather than through a specialization of general-purpose microprocessors.

Amazingly, few of the fundamentals have changed since that time. For example, the AT&T DSP16xx architecture plainly shows its ancestry: its basic architectural organization is little different from the AT&T DSP1, first produced in 1979. Similarly, the Texas Instruments TMS320C5x family has much the same organization as the TMS32010, first produced in 1982. Although there have been countless incremental improvements, the only dramatic architectural change to have a broad impact on DSP processors in the last ten years was the development of floating-point DSPs. To date, their higher cost has prevented floating-point DSPs from making substantial inroads into embedded systems applications—the traditional, high-volume domain of fixed-point DSPs. However, floating-point DSPs are becoming less expensive and have begun to score design wins in some embedded applications. While we expect that floating-point DSPs will continue to improve their price-performance, we also expect that fixed-point processors will continue to dominate in the largest applications for the foreseeable future.

Despite the similarity of current DSPs to those of fifteen years ago, there are signs of increasing diversity and innovation in processor architectures. For example, the Green Core from Infinite Solutions uses a design philosophy that is substantially different from those of other DSPs. We expect to see an increasing variety of DSP processor architectures over the next several years, including some based on RISC technology, on very large instruction word (VLIW) technology, and hybrid DSP microprocessors.

Market Challenges

The main market for DSPs has always been embedded systems. Voiceband data modems, for example, were an early application of DSPs and continue to be one of the most important. Other applications with a long history of DSP use include music, speech synthesis, and servo controllers. Speech compression applications have only recently begun using DSPs in high volume. This is primarily because recent wireless communications standards are better suited to implementation on DSP processors than were older and less aggressive compression standards, which were better suited to custom ASIC implementations.

To serve the embedded systems market well, DSP vendors must resist the temptation to excessively expand the feature sets provided by their devices. Most customers will be unwilling to pay the price for these features, in terms of overall system cost and power consumption.

A countervailing pressure, however, comes from the proliferation of DSPs into an ever-widening array of diverse applications, each with its own specialized needs. Several DSP vendors are successfully addressing this problem by producing a variety of processor variants based on the same DSP core. Some have introduced application-specific processors with special-purpose execution units (e.g., Viterbi decoders for wireless communications). Other vendors are offering DSP cores, which can become the basis of a customer's ASIC. This allows the system designer to customize the IC to very closely match the needs of an application. Reuse of a DSP core permits the core vendor and the customer to leverage their investments in development tools, software, and engineering know-how.

An additional market pressure comes from the increasing use of DSP processors in portable, battery-powered applications. Already most processor vendors offer 3.3 or 3.0 V versions of their DSPs to reduce power consumption. We expect to see supply voltages continue to drop and increasingly aggressive power management features being added to DSP processors in the next few years.

Competitive Challenges

The main competition for DSPs has always been custom analog, digital, or mixed circuitry. To meet this competition, DSPs have concentrated on speed and low cost, with memory capacity and I/O as secondary considerations. Given that the competition is custom circuitry, the assumption is usually that the DSP processor will be programmed in assembly language in an effort to obtain maximum efficiency.

Given the predominance of embedded systems applications, a possible new competitive threat to DSPs is high-level synthesis of ASICs. A design tool like Mentor Graphics' DSP Station could, in concept, replace DSPs with fully customized devices, while providing the designer with a straightforward programming interface. This threat, however, has thus far failed to materialize for several reasons. First, the technical problems are daunting, and current synthesis systems have demonstrated success only for narrow application areas. Second, the cost of fully customized devices may still be higher than that of programmable DSP solutions due to the higher volume production of DSPs.

Many manufacturers of high-performance general-purpose microprocessors believe that they, too, are a threat to DSPs. There is currently little evidence that this is true for the embedded systems market, except possibly for floating-point DSPs. While many microprocessors have acquired DSP-like features such as fast multipliers, their system cost is much too high for most embedded systems. Moreover, the high system cost is largely due to features that are irrelevant in embedded systems. These include elaborate floating-point hardware with sophisticated handling of exceptions, cache management (which yields a nondeterministic speed gain not useful for most hard-real-time applications), and hardware support for virtual memory.

It can be argued, however, that conventional microprocessors might eventually displace DSPs in certain types of systems, especially where a hardware platform containing a powerful general-purpose microprocessor is already established in a market. Intel's native signal processing initiative envisions PC multimedia functions such as music synthesis and speech compression implemented in software running on the main CPU of a personal computer, doing away with the need for a separate DSP processor. Currently, host processors are capable only of low-performance signal processing tasks, and manage such functions only at the expense of leaving few processor cycles available for other uses. However, one can anticipate sufficient improvement in the performance of microprocessors that this computational burden will become tolerable. At the same time, one can anticipate that, even as general-purpose microprocessors improve, communications technology will also advance, making ever-greater computational demands on hardware. For example, voiceband data modems might be replaced by wireless links to base stations with direct digital connections into a high-speed network.

When assessing alternatives to DSPs such as general-purpose microprocessors, it is critical to evaluate comparable technology. It would be very misleading, for example, to compare the projected capabilities of a microprocessor that will become available in two years to a DSP that is in volume production today. It is also inappropriate to compare an $800 microprocessor to a $50 DSP without considering cost. For many applications, it is equally inappropriate to compare a 5 W device to a 100 mW device without considering power consumption. Today, when all of these factors are considered, DSPs are still the overwhelming winners for most embedded signal processing applications.

Technical Challenges

Overall, software is the clear weak spot in DSP processor technology. This includes development tools supporting DSP-based software and hardware design, as well as libraries of software functions and applications programs suitable for use in end products.

Most tools supporting DSP processor software development are created by the DSP processor manufacturers, rather than by third-party software houses. Most vendors offer the traditional trio of assembler, linker, and simulator, and nothing more. This set of tools is oriented toward painstaking, cycle-by-cycle optimization of programs, and the tools are often little different from the technology of the 1970s.

Given that most DSP processor software is developed in assembly language, the low quality of most existing assembly language development and debugging environments is surprising. A key technical challenge facing DSP processor vendors is development of sophisticated tools that enable extremely detailed, highly productive software optimization and debugging. Users need symbolic debugging, the ability to graphically display data, and tools supporting structured, modular program development and software reuse. Other critical tool needs include good profiling tools (including tools to help the programmer visualize profiling results), tools to detect possible undesired side effects of instruction sequences in highly pipelined processors, and tools to assist system designers in accurately estimating power consumption for an application. Application and function libraries are also key to increasing developer productivity and speeding the expansion of DSPs into new applications. We believe that these tools offer a promising path to faster development of more complex application programs.

Although most DSP vendors have developed C compilers for their processors, these compilers (especially those for fixed-point processors) often seem to be most useful as a marketing tool for the vendor rather than as a development tool for the programmer. Although today's compilers are better than those of even a few years ago, the fact remains that most developers of applications software for DSP processors do not use C. This reflects the extreme cost sensitivity of their applications and their focus on embedded systems, which often justify painstaking software optimization. We expect that compilers will continue to improve and their use will become more widespread, but this will take time and significant effort on the part of the DSP tool vendors.

DSP core technology is already being used by DSP vendors to offer their customers more customized configurations of their processors. Today, the customization is usually done by the DSP processor manufacturer. But the true potential of DSP core technology will be realized when system designers have the ability to create customized DSP processors for their own applications. Currently, however, the design tools needed to support user-customization are sorely lacking. DSP core vendors will need to develop much closer cooperation with CAD tool vendors to create design approaches and tools that will allow customized DSPs to achieve their potential. Several vendors of DSP cores (such as Texas Instruments and DSP Group) have taken tentative steps down this path by working with CAD tool vendors to make available full-functional processor models of their cores. Much remains to be done in this area, however.

Conclusion

We believe that DSP processors will become increasingly important in an expanding range of electronic products over the next several years, much as microprocessors and microcontrollers have over the past two decades. The diversity of these applications and the stringent performance, cost, and power consumption demands they make will spur an increased pace of processor architectural innovation and specialization. This specialization is both a strength and a weakness of DSP processors, positioning them between custom ASICs and general-purpose microprocessors in terms of performance and system hardware and software development complexity. While DSP processors will be threatened by these alternative technologies in some applications, we believe that DSPs will be the implementation approach of choice in some of the next decade's most important applications, including telecommunications systems of many kinds, advanced multimedia user interfaces, and flexible information appliances.

Appendix

Vendor Contact Information

The following table provides contact information for vendors of DSP processors mentioned in this book.

DSP Processor Vendors			
Vendor	**Address**	**Telephone, Fax, Electronic Mail**	**Processor Families**
Analog Devices, Inc.	1 Technology Way P. O. Box 9106 Norwood, MA 02062	(617) 461-3672 (617) 461-4258 - BBS Email: dsp_applications@analog.com http://www.analog.com	ADSP-21xx ADSP-21020 ADSP-2106x
AT&T Microelectronics	555 Union Boulevard Allentown, PA 18103	(800) 372-2447 (610) 712-4593 - Fax (610) 712-4979 - BBS	DSP16xx DSP32xx DSP32C
Butterfly DSP, Inc.	2401 S.E. 161st Ct., Suite A Vancouver, WA 98684	(360) 892-5597 (360) 892-0402 - Fax	LH9320 LH9124
Clarkspur Design, Inc.	12930 Saratoga Avenue Suite B9 Saratoga, CA 95070	(408) 253-3196 (408) 253-3198 - Fax Email: yagi@clarkspur.com http://www.clarkspur.com/clarkspur	CD2400 CD245x
DSP Group, Inc.	3120 Scott Boulevard Santa Clara, CA 95054	(408) 986-4300 (408) 986-4323 - Fax http://www.dspg.com	OakDSPCore PineDSPCore
IBM Microelectronics	3039 Cornwallis Drive Research Triangle Park, NC 27709	(800) 426-0181 (415) 855-4121 - Fax http://www.chips.ibm.com	MDSP2780 ("Mwave")
Infinite Solutions, Inc.	3333 Bowers Avenue Suite 280 Santa Clara, CA 95054	(408) 986-1686 (408) 986-1687 - Fax Email: info@infinitesolutions.com http://www.infinitesolutions.com	Green Core

DSP Processor Vendors (Continued)			
Vendor	**Address**	**Telephone, Fax, Electronic Mail**	**Processor Families**
Motorola, Inc.	MD OE214 6501 William Cannon Drive Austin, TX 78735	(800) 521-6274 (512) 891-2030 - Marketing (512) 891-3230 - Tech. Support (512) 891-3771 - BBS (512) 891-4665 - Fax http://www.mot.com	DSP5600x DSP561xx DSP96002
Motorola, Inc.	Attn: Rex Kiang MD OE216 6501 William Cannon Drive Austin, TX 78735	(512) 891-2429 (512) 891-8807 - Fax rex_kiang@oakqm3.sps.mot.com	MC68356
NEC Electronics, Inc.	475 Ellis Street P.O. Box 7241 Mountain View, CA 94039	(415) 965-6000 (415) 965-6776 - Tech. Support (415) 965-6437 - Fax (415) 965-6466 - BBS	μPD7701x
SGS-Thomson Microelectronics	2055 Gateway Place Suite 300 San Jose, CA 95110 7 Avenue Gallieni F-94243 Gentilly Cedex FRANCE	(408) 452-8585 (408) 452-1549 - Fax +33-1-47-40-75-75 +33-1-47-40-79-10 - Fax	D950-CORE
TCSI Corporation	2121 Allston Way Berkeley, CA 94704	(510) 649-3700 (510) 649-3500 - Fax Email: info@tcs.com http://www.tcs.com	Lode core
Tensleep Design, Inc.	3809 South 2nd Street Suite D100 Austin, TX 78704	(512) 447-5558 (512) 447-5565 - Fax Email: info@tensleep.com	A/DSCx21 cores
Texas Instruments, Inc.	13510 N. Central Exwy. Dallas, TX 75265	(713) 274-2320 - Tech. Support (713) 274-2324 - Fax (713) 274-2323 - BBS Email: 4389750@mcimail.com http://www.ti.com	TMS320C1x TMS320C2x TMS320C2xx TMS320C3x TMS320C4x TMS320C5x TMS320C54x TMS320C8x
3Soft Corporation	1001 Ridder Park Drive San Jose, CA 95131-2314	(408) 451-5670 (408) 451-5690 - Fax Email: sales@3soft.com	M320C25 core M320C50 core

DSP Processor Vendors (Continued)			
Vendor	**Address**	**Telephone, Fax, Electronic Mail**	**Processor Families**
Zilog, Inc.	210 East Hacienda Avenue Campbell, CA 95008	(408) 370-8000 (408) 370-8056 - Fax http://www.zilog.com	Z893xx Z894xx
Zoran Corporation	2041 Mission College Blvd. Santa Clara, CA 95054	(408) 986-1314 (408) 986-1240 - Fax Email: vlad@zoran.com	ZR3800x

References and Bibliography

[All85] Jonathan Allen, "Computer Architecture for Digital Signal Processing," *Proceedings of the IEEE*, Vol. 73, No. 5, May 1985: 852(22).

[Ana90] Analog Devices, DSP Division, *Digital Signal Processing Applications Using the ADSP-2100 Family* (Vol. 1). Edited by Amy Mar. Englewood Cliffs, New Jersey: Prentice Hall, 1990.

[Ana95] Analog Devices, DSP Division, *Digital Signal Processing Applications Using the ADSP-2100 Family* (Vol. 2). Edited by Jere Babst. Englewood Cliffs, New Jersey: Prentice Hall, 1995.

[BDT95] Berkeley Design Technology, Inc., *DSP Design Tools and Methodologies*. Fremont, California: Berkeley Design Technology, Inc., 1995.

[Ber95] Shaul Berger, "DSP Seen as Enabling Technology for PCS," *Electronic Engineering Times*, April 17, 1995: p. 56(1).

[Bin94a] Ashok Bindra, "Digital Signal Processing," *Electronic Engineering Times*, February 21, 1994: p. 59(1).

[Bin94b] Ashok Bindra, "Digital Signal Processing," *Electronic Engineering Times*, April 18, 1994: p. 41(2).

[Bin95] Ashok Bindra, "Digital Signal Processing," *Electronic Engineering Times*, January 23, 1995: p. 35(1).

[Bur95] Dave Bursky, "DSP ICs Penetrate Into Low-Cost Applications," *Electronic Design*, March 20, 1995: p. 51(4).

[Bur85] C. S. Burrus and T. W. Parks, *DFT/FFT and Convolution Algorithms: Theory and Implementation*, with TMS32010 programs by James F. Potts. New York: Wiley, 1985.

[Chi94] Jeff Child, "Higher Levels of Integration Come to DSPs," *Computer Design*, May 1994: p. 91(6).

[Cha90] Rulph Chassaing and Darrell W. Horning, *Digital Signal Processing with the TMS320C25*. New York: Wiley, 1990.

[Cha92] Rulph Chassaing, *Digital Signal Processing with C and the TMS320C30*. New York: Wiley, 1992.

[Dwi95] David R. Dwin, "DSPs Start to Take Over PC's Tasks," *Electronic Engineering Times,* January 23, 1995: p. 36(3).

[Els90] Mohamed El-Sharkawy, *Real Time Digital Processing Applications with Motorola's DSP56000 Family*; with appendices provided by the Applications Engineering Staff of Motorola's DSP Operation. Englewood Cliffs, New Jersey: Prentice Hall, 1990.

[IEE85] Institute of Electrical and Electronics Engineers, Inc., *IEEE Standard for Binary Floating-Point Arithmetic*. New York: Institute of Electrical and Electronics Engineers, Inc., 1985.

[IEE87] Institute of Electrical and Electronics Engineers, Inc., *IEEE Standard for Radix-Independent Floating-Point Arithmetic*. New York: Institute of Electrical and Electronics Engineers, Inc., 1987.

[Jon88] Douglas L. Jones and Thomas W. Parks, *A Digital Signal Processing Laboratory Using the TMS32010*. Englewood Cliffs, New Jersey: Prentice Hall, 1988.

[Lap94a] Phil Lapsley, "Coming to Grips with DSP Benchmarks," *Electronic Engineering Times,* February 21, 1994: p. 62(1).

[Lap94b] Phil Lapsley and Jeff Bier, "DSP Cores Bring New Levels of Integration," *Microprocessor Report,* Vol. 8, No. 10, August 1, 1994: p. 1(4).

[Lap95] Phil Lapsley, "NSP Shows Promise on Pentium, PowerPC," *Microprocessor Report,* Vol. 9, No. 6, May 8, 1995: p. 11(5).

[Lee88] Edward A. Lee, "Programmable DSP Architectures: Part I," *IEEE ASSP Magazine,* October 1988: pp. 4-19.

[Lee89] Edward A. Lee, "Programmable DSP Architectures: Part II," *IEEE ASSP Magazine,* January 1989: pp. 4-14.

[Lev96] Markus Levy and Anne Coyle, "EDN's 1996 DSP Chip Directory," *EDN,* March 1, 1996: p. 40(27).

[Man88] M. Morris Mano, *Computer Engineering Hardware Design*. Englewood Cliffs, New Jersey: Prentice-Hall, 1988.

[Ohr93] Stephan Ohr, "New Applications Driving Dedicated DSP Processors," *Computer Design*, Vol. 31, No. 5, May 1993: p. 83(11).

[Opp89] A. Oppenheim and R. Schafer, *Discrete-Time Signal Processing*. Englewood Cliffs, New Jersey: Prentice-Hall, 1989.

[She88] David Shear, "EDN's DSP Benchmarks," *EDN*, September 29, 1988: pp. 126-148.

[Sie82] Daniel P. Sieworek, C. Gordon Bell, and Allen Newell, *Computer Structures: Principles and Examples*. New York: McGraw-Hill, 1982.

[Str95] Will Strauss, *DSP Strategies for the 90s, the Mid-Decade Outlook*. Tempe, Arizona: Forward Concepts, 1995.

[Tuc95] Barbara Tuck, "DSP Cores Provide Smaller, Cheaper, Faster, Lower Power Systems," *Computer Design,* Vol. 34, No. 7, July 1995: p. 50(2).

[Was82] Shlomo Waser and Michael J. Flynn, *Introduction to Arithmetic for Digital Systems Designers*. New York: Holt, Rinehart and Winston, 1982.

[Wei95] Ray Weiss, "Designers Go Shopping as DSP Prices Start Dropping," *Computer Design*, December 1995: p. 69.

Glossary

Absolute addressing — *See* Memory-direct addressing.

Accumulator — A register used to hold the output of the ALU or multiply-accumulate unit. On fixed-point processors, accumulators are usually at least twice as wide as the processor's basic data word width (in bits) and may be wider. DSP processors typically have from one to four accumulators.

A/D converter — (Analog-to-digital converter.) A circuit that converts an analog voltage into a numeric (digital) representation.

Address match breakpoint — A debugging feature provided by in-circuit emulators and instruction set simulators. The emulator or simulator halts processor execution when the processor attempts to access program or data from a specified memory address.

ADPCM — (Adaptive differential pulse code modulation.) An audio compression technique.

A/DSCx21 cores — A family of 16-bit fixed-point DSP cores from Tensleep Design, Inc.

ADSP-21xx — A family of 16-bit fixed-point DSP processors from Analog Devices, Inc.

ADSP-21020 — A 32-bit floating-point DSP processor from Analog Devices, Inc.

ADSP-2106x — A family of 32-bit floating-point DSP processors from Analog Devices, Inc., notable for their communications ports and large on-chip memories.

AE (Applications engineer.) An engineer employed by a processor vendor responsible for assisting customers with implementing their applications on that vendor's processors.

A-law A European encoding standard for digital representation of speech signals. Nonuniform quantization levels are used to achieve the effect of compressing the dynamic range prior to quantization. *See also* Companding, μ-law.

ALU (Arithmetic/logic unit.) An execution unit on a processor responsible for arithmetic (add, subtract, shift, and so on) and logic (*and*, *or*, *not*, *exclusive-or*) operations.

ANSI (American National Standards Institute.) A standards-setting body. Among other standards, ANSI has defined a standard for the C programming language.

ASIC (Application-specific integrated circuit.) An integrated circuit intended for use in a particular product or set of products, designed by the users of the IC.

Assembly statement in-lining The inclusion of an assembly language statement (or statements) in assembly code generated from a high-level language. Assembly statement in-lining is a technique commonly used with DSP processor high-level language compilers to improve performance.

ASSP (Application-specific standard product.) An ASSP is different from an ASIC in that it is intended to be used in a range of products within a particular application field, whereas an ASIC is often used only in a single product. *See also* ASIC, FASIC.

ATPG (Automatic test pattern generation.) The automated generation of test data (often called test vectors or test patterns) to be used for production testing of integrated circuits or other hardware components.

BBS (Bulletin board system.) Many DSP processor vendors provide BBSs accessible by modem that hold source code and application notes for their processors.

Big-endian A term used to describe the ordering of bytes within a multibyte data word. In big-endian ordering, bytes within a multibyte word are arranged most-significant byte first. *See also* Little-endian.

Biquad filter A second-order digital filter commonly used in signal processing. Biquads are often used as building blocks in higher-order digital filters.

Bit field manipulation Logical or bit operations applied to a group of bits at a time. Bit field manipulation is a key operation in error control coding and decoding.

Bit I/O port An I/O port in which each bit is individually configurable to be an input or an output and in which each bit can be independently read or written.

Bit-reversed addressing An addressing mode in which the order of the bits used to form a memory address is reversed. This simplifies reading the output from radix-2 fast Fourier transform algorithms, which produce their results in a scrambled order.

Block repeat A hardware looping construct in which a block of instructions is repeated a number of times.

Boundary scan A facility provided by some integrated circuits that allows the values on the IC's pins to be interrogated or driven to specified logic levels through the use of a special serial test port on the device. This is useful for testing the interconnections between ICs on a printed circuit board.

Bounding box The boundary of a circuit component, such as a DSP processor, ASIC, standard cell, or DSP core. The bounding-box definition includes a list of all signals and associated timing and electrical properties.

Branch A change in a processor's flow of execution to continue execution at a new address.

Bubble A delay in instruction execution caused by a pipeline conflict. So-called because when depicted graphically the unused pipeline slot appears to "bubble up" toward the execute stage of the pipeline.

Bus A shared electrical connection that moderates access to a resource, such as a memory shared among several hardware subsystems.

Bus-functional simulation model A partial simulation model of a programmable processor that models activity on the pins or bounding box only. It is not capable of simulating the execution of a program. *See also* Full-functional simulation model, Bounding box.

Butterfly An operation used in the computation of the fast Fourier transform. A butterfly computation involves a multiplication, addition, and subtraction of complex numbers.

CAD Computer-aided design.

CAE Computer-aided engineering.

Cascade of biquads An implementation for IIR filters where the transfer function is factored into second-order terms which are then implemented as a chain of biquad filters.

CD2400 A 16-bit fixed-point DSP core from Clarkspur Design, Inc.

CD2450 A fixed-point DSP core with configurable data word widths from Clarkspur Design, Inc.

cDSP (Configurable DSP.) Texas Instruments' term for DSP core-based ASICs.

CELP (Codebook excited linear prediction.) A speech coding technique. CELP usually refers to the CELP algorithm specified in USFS 1016 which compresses speech to 4800 bits/s. *See also* USFS 1016.

Circular addressing *See* Modulo addressing.

Circular buffer A region of memory used as a buffer that appears to wrap around. Circular buffers are typically implemented in software on conventional processors and via modulo addressing on DSPs.

Clock cycle The time required for one cycle of the processor's master clock. *See also* Instruction cycle.

Clock divider A circuit that reduces the frequency of a processor's master clock. Programmable clock dividers allow the programmer to slow down the processor's clock during times when full-speed operation is not needed, thus reducing power consumption.

Clock doubler A frequency synthesizer circuit that allows an input clock with frequency of one-half of the processor's desired master clock frequency to be used to generate the master clock. Both the Texas Instruments TMS320C5x family and the Analog Devices ADSP-2171 feature on-chip clock doublers. *See also* Phase-locked loop.

CMOS

(Complementary metal-oxide semiconductor.) A semiconductor technology that results in lower power consumption than other technologies. Most IC technology used today is CMOS.

Codec

(Coder-decoder.) An A/D and D/A converter for speech or telephony applications. The term "codec" carries with it an implication that the samples produced and consumed by the device are encoded using companding, but the term is not always used in this strict a fashion.

COFF

(Common object file format.) An object file format developed by AT&T. Most assembly language tools for DSP processors use and generate COFF files.

Companding

Short for compressing-expanding, a technique for reducing the dynamic range of audio signals and then later expanding it again. Companding uses a nonuniform quantization scheme that features finer quantization intervals at lower signal levels (where the input signal probabilistically spends more time) to achieve a higher signal-to-noise ratio in most cases. Companding is used in A-law and μ-law codecs.

Conflict wait state

A wait state (defined below) inserted due to contention for a resource. Similar to a pipeline interlock, except that a pipeline interlock is usually attributable to contention for a resource in the processor's core, whereas a conflict wait state may be attributable to contention for a resource outside of the core, such as an external memory interface.

Convergent rounding

A rounding technique used to avoid the bias inherent in the conventional round-to-nearest approach. This technique works by attempting to randomize rounding behavior in the case where the input value to be rounded lies exactly halfway between two output values. In half of these cases the value is rounded up, and in the other half of the cases it is rounded down.

Convolutional encoding

An error control coding technique used to encode bits before their transmission over a noisy channel. Used in modems and digital cellular telephony. Convolutional encoding is usually decoded via the Viterbi algorithm (see below).

Core

Refers to the central execution units of a processor, excluding such things as memory and peripherals. In many cases, DSP processor manufacturers use a common core with different combinations of memory and peripherals to create a family of processors with the same architecture. Some vendors provide DSP cores that their customers can use to create their own customized application-specific ICs.

COSSAP A block-diagram-based DSP design, simulation, and implementation environment from Synopsys, Inc.

CPP (C preprocessor.) A preprocessor for the C programming language. CPP is responsible for expanding macros, filtering out comments, and resolving conditional compilation directives. It is automatically invoked by the compiler. Some DSP assemblers use CPP to implement macros.

CQFP (Ceramic quad flat pack.) A type of IC packaging.

Cycle *See* Instruction cycle or Clock cycle.

Cycle stealing Delaying an operation to allow access to processor resources for another operation, such as DMA.

D950-CORE A 16-bit fixed-point DSP core from SGS-Thomson Microelectronics.

D/A converter (Digital-to-analog converter.) A circuit that outputs an analog voltage given a numeric (digital) representation of the voltage as input.

Data path A collection of execution units (adder, multiplier, shifter, and so on) that process data. A processor's data path determines the mathematical operations possible on that processor.

Data-stationary An instruction set design for pipelined processors where an instruction specifies the actions that should be performed on a set of data values, even if these actions are distributed over several instruction cycles and coincide with actions specified in other instructions. The AT&T DSP32C and DSP321xx are good examples of processors that use a data-stationary approach to programming. Contrast with time-stationary.

DAU (Data arithmetic unit.) AT&T DSP32C/DSP3210 and DSP16xx.

Debugger A front-end program that provides the user interface and much of the functionality of an emulator or instruction set simulator.

Delay line A buffer used to store a fixed number of past samples. Delay lines are used to implement both FIR and IIR filters.

Delayed branch A branch instruction where the branch actually occurs later than the lexical appearance of the instruction. In other words, one or more instructions appearing after the branch in the program are executed before the branch is executed.

Die A single integrated circuit as a portion of a silicon wafer.

DMA (Direct memory access.) A mechanism by which an external device or peripheral can access the processor's memory to transfer data without the processor having to execute data movement instructions.

DRAM (Dynamic random access memory.) DRAM provides greater memory densities and is cheaper than SRAM (see below), but is also slower and requires external circuitry to refresh it periodically. Some DSP processors provide on-chip DRAM interfaces.

DSP (1) Digital signal processing. (2) Digital signal processor.

DSP16xx A family of 16-bit fixed-point DSP processors from AT&T Microelectronics.

DSP32xx A family of 32-bit floating-point DSP processors from AT&T Microelectronics.

DSP32C A family of 32-bit floating-point DSP processors from AT&T Microelectronics.

DSP5600x A family of 24-bit fixed-point DSP processors from Motorola, Inc.

DSP561xx A family of 16-bit fixed-point DSP processors from Motorola, Inc.

DSP96002 A 32-bit floating-point DSP processor from Motorola, Inc.

DSPx A DSP-related conference and trade show.

DSP Station A DSP design, simulation, and implementation environment from Mentor Graphics Corporation.

Embedded system A system containing a processor (for example, a digital signal processor or a general-purpose microprocessor) wherein the processor is not generally reprogrammable by the end user. For example, a modem containing a DSP processor is an embedded system. A personal computer is not.

EPROM (Erasable programmable read-only memory.) An EPROM can be erased and reprogrammed by the user multiple times. *See also* PROM, ROM.

Event counter In the context of in-circuit emulators, an event counter counts the number of times a user-specified event occurs while the processor is executing. An event may consist of, for example, access to specified program or data memory addresses, a branch taken by a program, or an external interrupt. Not all in-circuit emulators provide event counters.

Exception An unplanned-for event that results from a software operation, such as division by zero.

Exponent A part of the representation of a floating-point number. *See also* Floating-point.

Externally-requested wait state Wait states that are requested by an external device. *See also* Wait state.

FAE (Field application engineer.) An application engineer who is based in one of a vendor's field offices.

FASIC (Function- and application-specific integrated circuit.) An integrated circuit that performs a specialized, high-level function (e.g., speech coding, image compression) that is sold off-the-shelf for use in the products of different companies. FASICs are sometimes referred to as ASSPs—application-specific standard products. *See also* ASIC, ASSP.

Fast interrupt An interrupt where the service routine can execute only one or two instructions but that offers reduced interrupt latency. Fast interrupts are typically used to quickly move data from a peripheral to a memory location or vice versa.

Feature size An overall indicator of the density of an IC fabrication process. It usually refers to the minimum size of one particular kind of silicon structure or "feature," specifically the minimum length of the "channel," or active region of a MOS transistor. The sizes of other structures on the IC are usually roughly proportional to the minimum transistor channel length. A smaller feature size translates into a smaller chip.

FFT (Fast Fourier transform.) A computationally efficient method of estimating the frequency spectrum of a signal. The FFT algorithm is widely used in DSP systems.

Field-programmable gate array A programmable logic chip having a high density of gates.

FIFO (First-in, first-out.) A type of buffer arrangement wherein the first sample stored in the buffer is the first one to be retrieved. *See also* Circular buffer.

FIR (Finite impulse response.) A category of digital filters. As compared to the other category, IIR filters, FIR filters are generally more expensive to implement, but offer several attractive design characteristics. *See also* IIR.

Fixed-point Pertaining to an arithmetic system where each number is represented using a fixed number of digits, and of these, a fixed subset specifies the integer part, with the remaining subset specifying the fractional part. Contrast with floating-point.

Floating-point Pertaining to an arithmetic system for representing integers or fractions where each number is represented by three fixed-point numbers, one specifying the sign, another the mantissa, and the third the exponent. The value of the number being represented is equal to the mantissa times the base (usually 2) raised to the power given by the exponent.

Full-functional simulation model In the context of commodity processor simulation models, a simulation model that models the processor internals and the activities on the bounding box of the processor. *See also* Bus-functional simulation model.

G.711 An ITU-T standard for encoding/decoding of audio signals. The standard compresses 16-bit samples of audio sampled at 8 kHz (128 kbits/s) to 64 kbits/s. The standard specifies two forms of companding: μ-law (used in North America and Japan) and A-law (used in Europe).

G.721 An ITU-T standard for speech encoding/decoding. The standard is based on ADPCM and compresses speech to 32 kbits/s. It is a subset of the G.723 algorithm. *See also* ADPCM, G.723.

G.722 An ITU-T standard for audio encoding/decoding. The standard is based on subband ADPCM and compresses a 7 kHz bandwidth audio signal to 64 kbits/s. *See also* ADPCM.

G.723 An ITU-T standard for speech encoding/decoding. The standard is based on ADPCM and compresses speech to 24, 32, or 40 kbits/s depending on the quality level desired. G.721 is G.723 running at 32 kbits/s. *See also* ADPCM, G.721.

G.728 The ITU-T standard for low-delay CELP (see above), a speech compression technique. G.728 compresses a 4 kHz audio bandwidth speech signal into a 16 kbit/s bit stream.

Gate array A digital integrated circuit consisting primarily of a regular array of cells. The final few steps of the fabrication process add customer-specified metal interconnection layers that both define the logic function of each cell and the interconnect of the cells. Because most of the fabrication steps are identical regardless of the application of the gate array, significant economies of scale are possible.

GCC (GNU C compiler.) A C compiler developed by the Free Software Foundation. GCC forms the basis of many C compilers for DSP processors.

GDB (GNU debugger.) A C-language source level debugger for use with GCC. *See also* GCC.

GNU (Gnu's not UNIX.) The name given by the Free Software Foundation to UNIX-like programs developed independently of AT&T or U.C. Berkeley and protected by a copyright agreement requiring free distribution of source and object code for original GNU software and derivative works.

Green Core A 16-bit fixed-point DSP core from Infinite Solutions, Inc.

GSM (Global System for Mobile Communications.) The standard specifying the pan-European digital cellular telephone network installed in the early 1990s. GSM also sometimes refers to the GSM standard speech coder, which compresses speech to 13.2 kbits/s. *See also* TDMA.

Guard bits Extra bits in an accumulator used to prevent overflow during accumulation operations. Most DSP processors provide from four to eight guard bits in their accumulators.

Hardware loop
A programming construct in which one or more instructions are repeated under the control of specialized hardware that minimizes the time overhead for the repetition.

Hardware stack
A push-down stack implemented in hardware. This facilitates subroutine calls and interrupt servicing. In DSP processors, these are usually small, which limits the depth of subroutine calls or interrupts.

Harvard architecture
A processor architecture with two separate banks of memory. The processor fetches instructions from one memory and data from the other. Most DSP processors are based on variants of the basic Harvard architecture.

Host interface (or port)
A specialized parallel port on a DSP intended to interface easily to a host processor. In addition to data transfer, some host interfaces allow the host processor to force the DSP to execute interrupt service routines, which can be useful for control.

Host processor
A general-purpose computer or microprocessor. Depending upon the context, a host computer could be a PC or workstation or might be a microprocessor used for control functions in an embedded system.

I^2C bus
Inter-integrated circuit bus, a synchronous serial protocol used to connect integrated circuits. The I^2C protocol was designed and promoted by Philips.

I^2S
A synchronous serial protocol developed by Philips and used for transferring digital audio signals between integrated circuits or between systems.

ICASSP
(International Conference on Acoustics, Speech, and Signal Processing.) One of the main technical conferences in the DSP field.

ICE
(In-circuit emulator.) A common tool for the development of microprocessor-based systems. An ICE usually consists of an adaptor, which takes the place of the processor in the target system, interface and control electronics, and software running on a host computer. Using an ICE, the engineer can interactively monitor and control the execution of the processor while it runs inside the target system. Many recently designed processors include ICE-like capabilities on-chip, with a serial port for host platform access.

IEEE standard 754 An IEEE standard for floating-point arithmetic. A number of DSP processors, including the Analog Devices ADSP-210xx and Motorola DSP96002, support IEEE-754 arithmetic.

IEEE standard P1149.1 An IEEE standard for boundary-scan testing of integrated circuits. A small serial port conforming to this standard is frequently used on DSP processors to access on-chip debugging facilities. IEEE standard P1149.1 is commonly called JTAG.

IIR (Infinite impulse response.) A category of digital filters. As compared to FIR filters, IIR filters generally require less computation to achieve comparable results, but sacrifice certain design characteristics which are often desirable. *See also* FIR.

Immediate addressing An addressing mode wherein the value to be used is specified as part of the instruction word. For example, "MOV #3,R0," which moves the value 3 into register R0, uses immediate addressing.

Instruction cycle The time required to execute the fastest instruction on a processor. *See also* Clock cycle.

Instruction set simulator A program that simulates the execution of programs on a specific processor. Instruction set simulators provide a software view of the processor: that is, they display program instructions, registers, memory, and flags, and allow the user to manipulate register and memory contents.

Interlock The delay introduced by an interlocking pipeline (see below) during a resource conflict.

Interlocking pipeline A pipeline architecture in which instructions that cause contention for resources are delayed by some number of instruction cycles. The Texas Instruments TMS320C3x, TMS320C4x, and TMS320C5x make heavy use of interlocking in their pipelines.

Interrupt An event that causes the processor to suspend execution of its current program and begin execution elsewhere in memory.

Interrupt latency The maximum amount of time from the assertion of an interrupt line to the execution of the first word of the interrupt's service routine, assuming that the processor is in an interruptible state.

I/O Input/output.

IS-54 A standard for U.S. digital cellular telephony.

IS-95 A standard for U.S. digital cellular telephony. IS-95 uses CDMA.

IS-136 A standard for digital cellular telephony. Also known as IS-54 revision C.

JEDEC Joint Electron Device Engineering Council.

Joule A unit of energy. One joule is the amount of energy used by a device consuming one watt of power in one second.

JTAG The informal name for IEEE standard P1149.1 (see above). JTAG stands for "Joint Test Action Group," the group that defined the standard.

Kernel (1) Software (such as an operating system) that provides services to other programs. (2) A small portion of code that forms the heart of an algorithm.

Linker A program that combines separate object code modules into a single object code module and resolves cross references between modules.

Little-endian A term used to describe the ordering of bytes within a multibyte data word. In little-endian ordering, bytes within a multibyte word are arranged least-significant byte first. *See also* Big-endian.

Lode core A 16-bit fixed-point DSP core from TCSI Corporation.

Loop unrolling A programming or compiler strategy whereby instruction sequences that are iterated are represented sequentially as separate instructions rather than within a loop. The overhead of looping is avoided. Loop unrolling is usually practical only for small loops.

Low voltage Pertaining to the use of less than the standard 5 V for digital logic. This is usually done to conserve power, but can also better match a given battery technology.

LPC (Linear predictive coding.) A speech coding and analysis technique.

LPC-10/LPC-10E A speech coder based on an LPC algorithm that compresses speech to 2400 bits/s. *See also* LPC, USFS 1015.

M320C25 core A 16-bit fixed-point DSP core from 3Soft Corporation.

M320C50 core A 16-bit fixed-point DSP core from 3Soft Corporation.

MAC *See* Multiply-accumulate.

Mantissa A part of the representation of a floating-point number. *See also* Floating-point.

Master clock The highest frequency clock signal used within a processor. The master clock is typically between one and four times the instruction execution rate of the processor.

MC68356 A single-chip multiprocessor containing both a DSP56002 and MC68000 processor from Motorola, Inc.

MCM (Multichip module.) A packaging technology that mounts integrated circuits (the dies themselves) directly on a substrate that interconnects them.

MDSP2780 A 16-bit fixed-point DSP with a 24-bit instruction word from IBM Microelectronics.

Memory-direct addressing An addressing mode where the address is specified as a constant that forms part of the instruction. For example, "MOV X:1234,X0" moves the contents of X memory location 1234 into register X0. *See also* Immediate addressing, Register-direct addressing, Register-indirect addressing.

Meta assembly language Assembly language that is parameterized with conditional constructs and variables, and does not contain direct specifications of processor registers. Meta assembly language is used in some code generation tools to allow more efficient code to be generated.

Micron A unit of length equal to 10^{-6} m. Integrated circuit feature sizes are usually specified in microns, and typical sizes range from 0.5 to 2.0 µm. *See also* Feature size. Abbreviated "µm."

Mil One one-thousandth of an inch. A unit sometimes used for describing integrated circuit die sizes.

Modifier register A register used in the computation of addresses. Some vendors use this term to refer to a register that contains a value to be added to an address register after an access is performed with that address register (post-incrementing). Other vendors use it to refer to a register that is used to configure a processor's address generation unit for a special addressing mode, such as modulo addressing or bit-reversed addressing.

Modulo addressing An addressing mode where post-increments are done using modulo arithmetic. This is used to implement circular buffers.

MOS (Metal-oxide semiconductor.) MOS is the silicon fabrication process used to implement most programmable DSP chips. The name refers to the three layers making up a transistor. The most common type of MOS used in digital circuits is CMOS, or complementary metal oxide semiconductor.

MQFP (Metal quad flat pack.) A type of IC package.

μ-law ("mu-law") An encoding standard for digital representation of speech signals. Nonuniform quantization levels are used to achieve the effect of compressing the dynamic range prior to quantization. *See also* Companding, A-law.

μPD7701x A family of 16-bit fixed-point DSPs with 32-bit instructions from NEC Electronics, Inc.

Multiply-accumulate The dominant operation in many DSP applications, where operands are multiplied and added to the contents of an accumulator register. Frequently abbreviated to "MAC."

Mwave A multimedia platform for the IBM PC developed by IBM Microelectronics. The Mwave multimedia architecture consists of the IBM Mwave DSP, a real-time operating system called Mwave O/S, software that allows IBM PC applications to communicate with DSP application code, and DSP applications code to implement various multimedia functions.

NaN (Not a number.) The IEEE standard 754 specifies that floating-point processors should reserve a special representation in their numeric formats to indicate that a register or memory location does not contain a valid number. This representation is referred to as NaN.

Nestable interrupts Interrupts whose service routines can be interrupted.

Normalization The detection and elimination of redundant sign bits in a fixed-point data word. Normalization is heavily used in block floating-point arithmetic.

Numerical C A GCC-based C compiler from Analog Devices, Inc. for its ADSP-210xx DSPs that implements a number of the extensions discussed by the Numerical C Extensions Group, in particular, complex arithmetic and iterators.

OakDSPCore A 16-bit fixed-point DSP core from DSP Group, Inc.

Object code Binary instructions and data for use on a programmable processor. Object code is usually produced by an assembler and is often "relocatable," meaning that it does not contain absolute references to particular memory locations.

Off-core A resource (such as memory or a peripheral) that is not contained within the actual processor core.

OnCE (On-chip emulation port.) A serial debugging port found on Motorola DSP processors.

On-core A resource (such as memory or a peripheral) that is contained within the actual processor core.

Operand-related/ operand-unrelated parallel move *See* Parallel move.

Orthogonal instruction set An instruction set where separate components (like the arithmetic operations, operand specifications, addressing modes, and parallel moves) of an instruction are encoded independently in separate fields of the instruction word. The principal consequence is that choosing one component (such as the arithmetic operation) does not constrain the other components (such as the addressing modes). This makes programming easier.

OTP One-time programmable.

Overflow A situation that occurs when the result of a mathematical operation (typically an add or subtract) requires more bits than are available in the register to which it is to be stored. Typical strategies for dealing with overflow are to saturate (meaning to limit the value at the most positive or negative representable value) or to wrap-around (meaning to store only the least significant bits that fit in the destination register).

Parallel move A movement of data that is carried out in parallel with the execution of an instruction. DSP processors typically provide the ability to move two data values in parallel with executing an instruction, although the number of instructions that support parallel moves may be limited.

PGA (Pin grid array.) A type of integrated circuit package. The external connections are made available on metal pins arranged in a grid.

Phase-locked loop A feedback system in which an oscillator tracks a periodic input signal. There are many uses for phase-locked loops, including timing recovery in modems and generation of an on-chip master clock at a higher frequency than an off-chip system clock.

PineDSPCore A 16-point fixed-point DSP core from DSP Group, Inc.

Pipeline An organization of computational hardware in which different stages of the execution of an instruction proceed in parallel for different instructions.

PQFP (Plastic quad flat pack.) A type of integrated circuit package.

Prioritized interrupts A scheme used to determine which of several simultaneous interrupts is serviced. Interrupts with higher priority are serviced first.

Profiling The process of determining the amount of time a processor spends in different sections of a program. The results of profiling are useful during the process of optimizing software for execution speed.

Programmable DSP An integrated circuit implementing a programmable processor suitable for signal processing.

Programmed wait state A wait state that is automatically generated by the processor when accessing certain ranges of external memory. Most processors allow the number of wait states to be configured by the programmer.

PROM (Programmable read-only memory.) PROM memory can be programmed once by the user after the chip has been fabricated. This is sometimes called one-time-programmable memory. *See also* EPROM, ROM.

PWM (Pulse width modulation.) PWM is used in some control applications. It is also sometimes used as an inexpensive way to implement a D/A converter.

QFP (Quad flat pack.) A type of integrated circuit package. ICs packaged in QFP packages are typically less expensive than the same IC in PGA packages.

Quick interrupt *See* Fast interrupt.

Real-time breakpoint A debugging feature provided by in-circuit emulators. The processor executes at full speed and execution is halted when a specified condition (called the breakpoint condition) evaluates true.

Real-time operating system (RTOS) An operating system that allows the developer to place an upper bound on the amount of time a process must wait to execute after a critical event occurs. Examples of real-time operating systems for DSPs include SPOX and Mwave O/S; UNIX is an example of a non-real-time operating system, in that programs may wait an indefinite amount of time before executing.

Register A circuit that holds a set of contiguous bits that are treated as a group. An accumulator is an example of a register.

Register-direct addressing An addressing mode where operands come from registers, and the registers are identified by constants in the instruction. For example, the instruction "ADD X0,Y0,A," which adds the contents of the X0 and Y0 registers and places their sum in the A register, uses register-direct addressing.

Register-indirect addressing An addressing mode in which the operand address is contained by a register, and the register is identified in the instruction. For example, the instruction "MOVE *R0,A," which moves the contents of the memory location whose address is stored in register R0 to register A, uses register-indirect addressing. In DSPs, the contents of the register is often modified before or after the address is extracted.

Relocatable code Object code that does not contain absolute memory addresses, but instead has symbolic references that a loader can resolve when it loads the program. This allows the program to be loaded into memory at any starting address. *See also* Object code.

Reverse-carry arithmetic
An alternative to bit-reversed addressing, where an increment of an address is done by adding a high-order bit and propagating the carry signal in the reverse direction, toward the low-order bit.

RISC
(Reduced instruction set computer.) A computer architecture with simple instructions that can be executed very quickly.

ROM
(Read-only memory.) ROM refers to mask-programmed ROM, meaning ROM whose memory contents are fixed when the chip is fabricated. *See also* PROM, EPROM.

Round-to-nearest
A rounding technique wherein numbers are rounded to the nearest representable number. A value that is equidistant between two representable numbers is always rounded up (or down, depending on the implementation). This introduces bias into calculations.

RTOS
(Real-time operating system.) *See* real-time operating system.

Saturation
A strategy for handling overflow in which the largest representable magnitude is used. Contrast with wrap-around.

Scan-based debugging/emulation
A debugging approach that uses dedicated hardware on a processor to debug the processor while it is operational and in-circuit within its target system.

Shadow register
A register that mirrors the contents of another register. This is useful, for example, for storing processor state during interrupt servicing. A shadow register can be thought of as a one-deep hardware stack that only supports a specific register.

Sleep mode
A power-conservation mode in which much of the hardware on the DSP is turned off.

SPI
(Serial peripheral interface.) A synchronous serial protocol used to connect integrated circuits.

SPOX
A real-time operating system for DSPs from Spectron Microsystems.

SQFP
(Shrunken quad flat pack.) A type of IC package.

SRAM
(Static random-access memory.) SRAM is often used in DSP systems because it is very fast and does not require periodic refreshing, as

DRAM does. However, SRAM is more expensive than DRAM and is not available in as high densities.

Standard cell A physical layout element in an IC foundry's library, conforming to certain standard characteristics, such as physical dimensions and electrical drive capability. A standard-cell integrated circuit is constructed by tiling together standard cells and interconnecting them.

Static When used to describe a processor, static means that the processor will run with an arbitrarily low frequency input clock and still function correctly, although more slowly. This is in contrast to a dynamic processor, which requires a minimum frequency input clock to function correctly. Because power consumption is proportional to clock frequency in CMOS circuitry, a static processor allows one to reduce power consumption by slowing or stopping the input clock.

Subroutine A unit of software that can be invoked from multiple locations in one or more other units of software to perform a specific operation or set of operations. Subroutines allow a programmer to avoid the need for repeatedly specifying often-used sequences of instructions in a program.

Subroutine call The action by which a processor transfers execution to a subroutine. At a minimum, this usually involves storing the current program counter value (so that execution can be resumed at the correct location when the subroutine completes) and executing a branch instruction.

T.4 An ITU-T standard for lossless compression and decompression of two-tone images based on Huffman coding. It is used in Group 3 facsimile machines.

Tap A fundamental section of an FIR filter, consisting of a coefficient, a multiplier, and a delay line stage.

Target system The end system or product in which a processor will be used.

TDMA (Time-division multiple access.) A multiple access method in which the channel is divided into multiple time slots, and only one user transmits in a given time slot.

TI Texas Instruments, Inc.

Time-stationary An instruction set design for pipelined processors in which an instruction specifies the operation performed by the various pipeline stages in one instruction cycle. The AT&T DSP16xx and all Motorola DSPs are good examples of this style. Contrast with data-stationary.

TMS320C1x A family of 16-bit fixed-point DSPs from Texas Instruments, Inc.

TMS320C2x A family of 16-bit fixed-point DSPs from Texas Instruments, Inc.

TMS320C2xx A family of 16-bit fixed-point DSPs from Texas Instruments, Inc.

TMS320C3x A family of 32-bit floating-point DSPs from Texas Instruments, Inc.

TMS320C4x A family of 32-bit floating-point DSPs from Texas Instruments, Inc.

TMS320C5x A family of 16-bit fixed-point DSPs from Texas Instruments, Inc. The TMS320C5x is the successor to the TMS320C2x family.

TMS320C54x A family of 16-bit fixed-point DSPs from Texas Instruments, Inc.

TMS320C80 A DSP processor from Texas Instruments, Inc. The TMS320C80 contains five processors on a single chip: a RISC-based "master processor" and four 16-bit, fixed-point DSP processors. The TMS320C80 is intended for image and video processing applications.

TQFP (Thin quad flat pack.) A type of integrated circuit package similar to, but thinner than, a plastic quad flat pack (PQFP; see above). TQFP packages are typically used in small, portable electronic systems, such as cellular telephones and pagers.

Two's complement The binary representation of numbers most commonly used in DSPs for fixed-point numbers.

USFS 1015 (United States Federal Standard 1015.) The standard specifying the LPC-10E speech coder. *See also* LPC.

USFS 1016 (United States Federal Standard 1016.) The standard specifying the CELP speech coder. *See also* CELP.

V.22bis An ITU-T standard for 2400 bit/s modems.

V.27 An ITU-T standard for 4800 and 2400 bit/s facsimile modems.

V.29 An ITU-T standard for 9600 and 7200 bit/s facsimile modems.

V.32 An ITU-T standard for 9600 bit/s modems.

V.32bis An ITU-T standard for 14,400 bit/s modems.

V.32terbo A protocol for 19,200 bit/s modems promulgated by AT&T before the V.34 standard was available.

V.34 An ITU-T standard for 28,800 bit/s modems.

V.42 An ITU-T standard for error correction.

V.42bis An ITU-T standard for data compression.

VCOS (Visible caching operating system.) (1) An architecture for PC multimedia formerly marketed by AT&T Microelectronics. (2) The real-time operating system used in the VCOS architecture.

Verilog HDL A hardware description language originally developed by Gateway Design Automation (now part of Cadence Design Systems, Inc.) for use with their digital hardware simulator. Verilog HDL is now a public standard, maintained by Open Verilog International. The language supports modeling of hardware at levels of abstraction ranging from gate level up to very abstract behavioral or performance models. Numerous companies market design entry, simulation, and synthesis tools that process Verilog HDL. *See also* VHDL.

VHDL (VHSIC hardware description language.) A hardware description language specified by IEEE standard 1076. The language supports modeling of hardware at levels of abstraction ranging from gate level to very abstract behavioral or performance models. Numerous companies market design entry, simulation, and synthesis tools that process VHDL. *See also* Verilog HDL.

VHSIC (Very high-speed integrated circuit.) A U.S. government program aimed at improving integrated circuit technology. The program is now defunct. The VHDL language resulted from this program.

Viterbi decoding (or Viterbi algorithm) A computationally efficient (but still relatively complex) mechanism for decoding a convolutionally encoded bit stream.

VLSI Very large scale integration.

VSELP (Vector sum excited linear prediction.) A speech coding technique used in the U.S. IS-54 digital cellular telephone system.

Wait state A delay inserted in an external memory access to give a slow peripheral or memory time to access its data.

Watt A unit of power. One watt is the power consumed by a device that uses one joule of energy in one second.

Wrap-around An overflow strategy in which overflow is ignored, and results are allowed to wrap around the ends of the range of representable numbers. The least significant bits (those that fit) are preserved.

Z893xx A family of 16-bit fixed-point DSPs from Zilog, Inc.

Z894xx A family of 16-bit fixed-point DSPs from Zilog, Inc.

ZR3800x A family of 20-bit fixed-point DSPs with 32-bit instruction words from Zoran Corporation.

Index

About the Authors

This book was prepared by the staff of Berkeley Design Technology, Inc. (BDTI), a firm founded in 1991 to make DSP technology more accessible to a wide range of product developers and to facilitate the commercialization of promising research technology. BDTI specializes in DSP technology evaluation and has produced a number of industry reports, including *Buyer's Guide to DSP Processors* and *DSP Design Tools and Methodologies*. (In fact, much of the material in this book is adapted from the introductory material in *Buyer's Guide to DSP Processors*.) BDTI also offers consulting services in the DSP field, ranging from technical evaluations of DSP processors and tools to customization and integration of EDA tools.

BDTI can be reached by telephone at +1 510 665-1600, by fax at +1 510 665-1680, by electronic mail at info@bdti.com, or via the World Wide Web at http://www.bdti.com.

Philip D. Lapsley is a founder of Berkeley Design Technology, Inc., where he is responsible for special projects. His technical interests include real-time DSP, DSP processor code generation and debugging, management of large software systems, network protocols, and computer security. He has worked at several research groups at the University of California at Berkeley, the NASA Ames Research Center, Teknekron Communication Systems, and the U. C. Berkeley Space Sciences Lab. Lapsley has also worked as an independent consultant in the field of real-time DSP, with emphasis on the interaction between real-time and non-real-time systems. At U. C. Berkeley, he cofounded the Experimental Computing Facility and served as its Director from 1986 to 1988. While a researcher in the DSP Design Group at U. C. Berkeley, Lapsley focused on DSP code generation and debugging, concentrating on the interface between programmable DSPs and host processors. This work culminated in the development of a debugger/monitor that allows users to monitor, control, and debug automatically generated DSP assembly code at the block-diagram level. He received both his B.S. degree with high honors and his M.S. degree in electrical engineering and computer sciences from the University of California at Berkeley.

Jeffrey C. Bier is a founder of Berkeley Design Technology, Inc., where he is responsible for general and technical management, research, and product development. His experience spans software, hardware, and design tool development for signal processing and control applications in commercial and research environments. Bier has held positions with Acuson Corporation, Hewlett-Packard Laboratories, Quinn & Feiner, the University of California at Berkeley, and elsewhere. He has implemented real-time signal processing systems using DSP processors from AT&T, Motorola, and Texas Instruments. While a researcher at U. C. Berkeley, Bier made significant contributions to the Gabriel DSP design project. He developed code generation and simulation software for multiprocessor DSP systems and was a key contributor to the development of a new class of high-efficiency multiprocessor architectures for DSP. In addition, he has developed several DSP ASICs. Bier has written numerous technical articles on topics including design tools, multiprocessor architectures, and simulation techniques. He earned his bachelor's degree with high honors from Princeton University. His master's degree is from the University of California at Berkeley.

Amit Shoham is a Senior DSP Engineer with Berkeley Design Technology, Inc., where he focuses primarily on benchmarking DSP processor performance and evaluating DSP design tools. His technical interests include digital audio and music synthesis. Prior to joining BDT, Mr. Shoham was at Silicon Graphics, where he developed diagnostics for digital audio hardware. He holds a bachelor's degree in computer systems engineering and a master's degree in electrical engineering, both from Stanford University.

Edward A. Lee is a professor in the Electrical Engineering and Computer Science Department at the University of California at Berkeley and a founder of Berkeley Design Technology, Inc. He has been codirector of the Ptolemy project (a system-level design and simulation project) at U. C. Berkeley since its inception in 1990 and he directed the Gabriel project before that. His research areas include parallel computation, architecture and software techniques for programmable DSPs, design environments for development of real-time software, and digital communication. He is a fellow of the IEEE and he was a recipient of a 1987 NSF Presidential Young Investigator award, an IBM faculty development award, the 1986 Sakrison prize at U. C. Berkeley for the best thesis in electrical engineering, and a paper award from the IEEE Signal Processing Society. He is coauthor of *Digital Communication*, with D. G. Messerschmitt, *Digital Signal Processing Experiments* with Alan Kamas, and numerous technical papers. His B.S. degree is from Yale University, his master's from MIT, and his Ph.D. from U. C. Berkeley. From 1979 to 1982 he was a member of the technical staff at Bell Telephone Laboratories, where he did extensive work with early programmable DSPs and exploratory work in voiceband data modem techniques and simultaneous voice and data transmission.

Printed in the United States
39820LVS00003B/217